Packet: Speed, More Speed and Applications

A Collection of Advanced Packet Methods and Activities from ARRL Publications and Other Sources

Compiled by Rich Roznoy, K1OF

Production Staff:
Paul Lappen
Joe Shea

Cover:
Design: Sue Fagan

Published by:
The American Radio Relay League
Newington, CT 06111 USA

Printed in the USA

Second Edition
First printing

ISBN: 0-87259-605-2

Foreword

The 2nd edition of *Packet: Speed, More Speed* is indeed "more." Picking up where the 1st edition left off, it provides, all in one convenient volume, an impressive compilation of articles and other information on advanced packet applications. The best news is that *there's no carryover from the 1st edition*: If you have a copy of the 1st edition, there's no duplication in this book.

Resourceful amateurs have used commercial technology developed during the past 30 years to make packet radio one of our hobby's most exciting modes. In the early days, receiving error free data was the principal goal. Today that is accomplished easily. Now the challenge is to have this error free data represent something more than plain text messages. With the new applications now available, data may represent maps, geographic locations, weather, DX stations and much more. A series of impressive new uses is possible due to the skill and generosity of Bob Bruninga, WB4APR, who developed the APRS (Automatic Packet Reporting System). He coupled a packet modem and a hand-held transceiver with a GPS (Global Positioning System), allowing us to track real-time events such as the path of a balloon as it drifts across the countryside. Mark and Keith Sproul, KB2ICI and WU2Z respectively, are developing add-on applications for APRS. Skywarn weather reporting, DX Cluster monitoring and high resolution map displays are just a few of these innovations.

As 1200 baud has become routine, innovators like Dale Heatherington, WA4DSY and TAPR (Tucson Amateur Packet Radio) are drawing on commercial technologies to give us a 50-fold increase in speed. If you, like many of us, don't have a high speed modem, you still can enjoy an exciting new use for your 1200-baud station. Fuji-OSCAR 29 offers us the first really user-friendly satellite BBS. You don't need any special equipment; a 1200-baud TNC used for your local BBS will work just fine.

As you can see, there are a lot of exciting new things to do on packet. This book provides you with the most complete single source of advanced packet information available. Let us know what you are doing and what material you would like to see in future editions. A handy feedback form is at the back for your convenience.

David Sumner, K1ZZ
Executive Vice President

Newington, Connecticut
March 1997

Contents

Chapter 1: Operating

Chapter 2: Theory/Design

Chapter 3: Networking

Chapter 4: Construction

Acknowledgments

This book is the product of many individuals. Our appreciation goes out to all who made contributions by supplying information and granting permission to use their material. A special thank you goes to Gloria Medcalf, KA5ZTX, author of *What is your TNC Doing?*, for sharing her "Radio-TNC Wiring Diagrams" with us.

About the
American Radio Relay League

The seed for Amateur Radio was planted in the 1890s, when Guglielmo Marconi began his experiments in wireless telegraphy. Soon he was joined by dozens, then hundreds, of others who were enthusiastic about sending and receiving messages through the air—some with a commercial interest, but others solely out of a love for this new communications medium. The United States government began licensing Amateur Radio operators in 1912.

By 1914, there were thousands of Amateur Radio operators—hams—in the United States. Hiram Percy Maxim, a leading Hartford, Connecticut, inventor and industrialist saw the need for an organization to band together this fledgling group of radio experimenters. In May 1914 he founded the American Radio Relay League (ARRL) to meet that need.

Today ARRL, with more than 170,000 members, is the largest organization of radio amateurs in the United States. The League is a not-for-profit organization that:

• promotes interest in Amateur Radio communications and experimentation

• represents US radio amateurs in legislative matters, and

• maintains fraternalism and a high standard of conduct among Amateur Radio operators.

At League headquarters in the Hartford suburb of Newington, the staff helps serve the needs of members. ARRL is also International Secretariat for the International Amateur Radio Union, which is made up of similar societies in more than 100 countries around the world.

ARRL publishes the monthly journal *QST*, as well as newsletters and many publications covering all aspects of Amateur Radio. Its headquarters station, W1AW, transmits bulletins of interest to radio amateurs and Morse code practice sessions. The League also coordinates an extensive field organization, which includes volunteers who provide technical information for radio amateurs and public-service activities. ARRL also represents US amateurs with the Federal Communications Commission and other government agencies in the US and abroad.

Membership in ARRL means much more than receiving *QST* each month. In addition to the services already described, ARRL offers membership services on a personal level, such as the ARRL Volunteer Examiner Coordinator Program and a QSL bureau.

Full ARRL membership (available only to licensed radio amateurs) gives you a voice in how the affairs of the organization are governed. League policy is set by a Board of Directors (one from each of 15 Divisions). Each year, half of the ARRL Board of Directors stands for election by the full members they represent. The day-to-day operation of ARRL HQ is managed by an Executive Vice President and a Chief Financial Officer.

No matter what aspect of Amateur Radio attracts you, ARRL membership is relevant and important. There would be no Amateur Radio as we know it today were it not for the ARRL. We would be happy to welcome you as a member! (An Amateur Radio license is not required for Associate Membership.) For more information about ARRL and answers to any questions you may have about Amateur Radio, write or call:

ARRL Educational Activities Dept
225 Main Street
Newington CT 06111-1494
(860) 594-0200

Prospective new amateurs call:
800-32-NEW HAM (800-326-3942)

E-mail: ead@arrl.org
Internet: http://www.arrl.org/

Chapter 1
Operating

GATEWAYS: Amateur Radio Meets the Internet

By Steve Ford, WB8IMY
Assistant Managing Editor
Internet: sford@arrl.org

I'm sitting at the keyboard of a friend's packet station. He's a relatively new packeteer, and I've been trying to convert him to the wonders of TCP/IP. That's the Amateur Radio version of the *Internet,* the global information superhighway.

It's been tough sledding, though. First we had to set up the software in his computer. Then I had to acquaint him with an entirely new way of navigating his local network.

"TCP/IP is all well and good, but what can I *do* with it?" he asked. "What happens after you go home to Connecticut and I'm left with this?"

"Well, we can chat and send electronic mail," I replied.

"Yeah, right. We tried that with the regular packet BBSs, remember? It took a week for my messages to reach you—when they reached you at all. Besides, you said my TCP/IP network was local. How could I possibly connect to you?"

This is *exactly* the moment I was waiting for. It was time to unveil my secret weapon. I tapped furiously on the keys and soon established a *telnet* session to a station on his network.

"What are you—"

"Sit tight. You'll see."

He peered closely at the screen as I entered:

telnet 44.88.4.35

I just instructed the local station to find a route to 44.88.4.35—my *home* TCP/IP address.

My friend shook his head. "You've asked that station to do the impossible. Our networks don't connect."

"Are you sure?" I said with a smile. "What if this station has a way to pass data to and from the Internet? What would happen then?"

His eyes widened. "A gateway!"

"Hmmm . . . could be!"

A thousand miles away, my packet station is faithfully monitoring 145.53 MHz. Another station belonging to Bill Lyman, N1NWP, is on the same frequency. N1NWP is also linked directly (by wire) to the Internet.

Sure enough, the link request finally arrives at N1NWP via the Internet. His computer "looks" at my IP address and knows exactly what to do. A connect request blasts out on 145.53 MHz. The *virtual circuit* is established and, halfway across the continent, I'm rewarded with the message shown in Figure 1

"Incredible!" my friend gasps. "So this is how the fabled gateways work. I thought they were just rumors. This is unbelievable!"

The Elusive Goal

As amateur packet radio has spread throughout the world, one goal has remained elusive: finding a fast, reliable way to transfer information between distant points. At one time we thought that we could build high-speed RF packet networks encircling the Earth, but this is not likely to happen anytime soon. The cost is outrageous and the required maintenance would be too much for most individuals or groups to bear.

We've been using HF packet to help bridge the gaps, but its performance leaves much to be desired. Other HF digital modes have improved the situation, but they're all at the mercy of the changeable nature of HF propagation. None of them offer the speed that even a 1200-bit/s VHF/UHF packet link can achieve.

Digital Amateur Radio satellites have been pressed into service to transfer packet mail. The satellites are reasonably dependable, but they are not *geostationary* (stationed at fixed points in the sky from our perspective on the ground). Instead, they orbit the Earth at relatively low altitudes. This means that packet stations have only a few opportunities each day to get data to and from the satellites.

It wasn't long before many hams began to consider *nonamateur* means to reach the same end. That's when they began taking a serious look at the Internet.

Opening the Gateways

Internet connections are as close as the nearest telephone lines. With the proper hardware and software, it isn't difficult to interface amateur packet radio stations to the Internet. Many of these become the so-called "gateways."

Some gateway stations are set up at

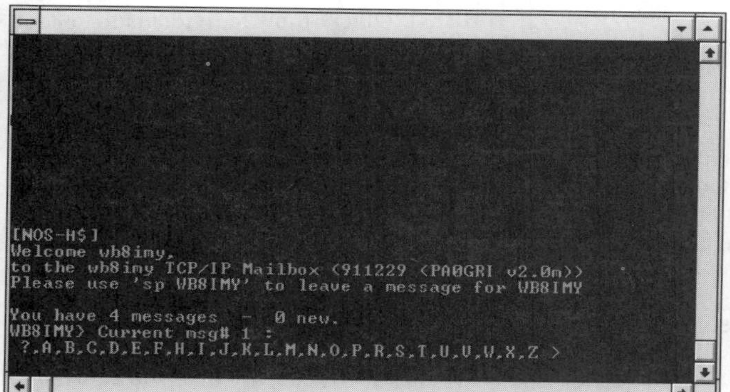

```
[NOS-H$]
Welcome wb8imy,
to the wb8imy TCP/IP Mailbox (911229 (PA0GRI v2.0m))
Please use 'sp WB8IMY' to leave a message for WB8IMY

You have 4 messages  -  0 new.
WB8IMY> Current msg# 1 :
 ?,A,B,C,D,E,F,H,I,J,K,L,M,N,O,P,R,S,T,U,V,W,X,Z >
```

Figure 1—By using an Internet gateway on my friend's TCP/IP network, I've connected to my home packet station—from more than 1,000 miles away!

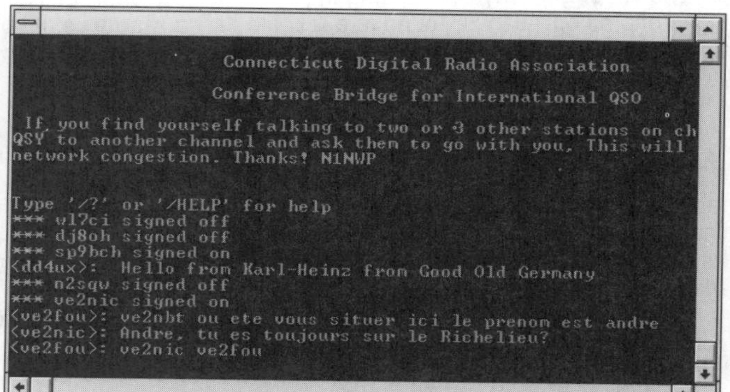

```
          Connecticut Digital Radio Association

        Conference Bridge for International QSO

 If you find yourself talking to two or 3 other stations on ch
QSY to another channel and ask them to go with you. This will
network congestion. Thanks! N1NWP


Type '/?' or '/HELP' for help
*** vl7ci signed off
*** dj8oh signed off
*** sp9bch signed on
<dd4ux>:  Hello from Karl-Heinz from Good Old Germany
*** n2sqw signed off
*** ve2nic signed on
<ve2fou>: ve2nbt ou ete vous situer ici le prenom est andre
<ve2nic>: Andre, tu es toujours sur le Richelieu?
<ve2fou>: ve2nic ve2fou
```

Figure 2—This is a brief glimpse of the international QSO conference bridge. It's especially hot and heavy during the evenings and on weekends.

colleges, universities or businesses. Because the Internet connections already exist, the people in charge allow the gateway operators to use them. Other gateway stations are operated by hams who have Internet access at their homes for personal or business use. (This is the case with the N1NWP gateway.)

Once these gateway stations transfer packet radio data to the Internet, the information moves at high speeds to almost anywhere in the world. Data also pours out of the Internet through the gateways and ends up on amateur packet networks.

Although it isn't a common-carrier service like the telephone companies, the reliability and speed of the Internet are impressive. You can think of the Internet as a multilane superhighway. The amateur packet network is likened to the slower surface streets. And what about the gateway stations? They're the entrance and exit ramps.

Internet gateway stations have been popping up all over the place in recent years. Hams are using them to act as links between packet networks. For example, a packet bulletin board system (PBBS) in New England may pass messages to the West Coast via an Internet gateway, rather than relying on the HF modes or satellites. This is often known as a *wormhole* link. Messages that took days to reach their final destinations arrive within hours or even minutes. Other hams tap the gateways and explore on their own, using the Internet as a packet version of the proverbial magic carpet.

TCP/IP

If you've been around packet radio long enough, there's a fair chance you've heard about *TCP/IP*. TCP/IP is an Amateur Radio adaptation of the protocols used on the Internet. Amateur TCP/IP networks (referred to collectively as the *AMPRNet*) have been growing throughout the nation. When you consider the similarity between amateur TCP/IP and the Internet, it's easy to understand why most gateway stations are on TCP/IP rather than "regular" (AX.25) packet.

You'll find TCP/IP activity primarily on 2 meters and 70 centimeters—usually on the same frequencies where you'd expect to hear regular packet. If you try to copy a TCP/IP signal with AX.25 protocol, however, you're likely to see gibberish on your monitor.

If you're already involved in packet, converting to TCP/IP is straightforward. All you need is the *NOS* software (in one of its many versions). Your terminal node controller (TNC) must be capable of operating in the *KISS* mode, but all TNCs manufactured in the past eight years or so include this feature.

Setting up *NOS* is difficult for some. You have to learn Unix-type jargon, for example, and change a number of files to adapt the software to the network as it exists in your area. To make matters worse, most versions of *NOS* are not especially friendly.

You can purchase books that will help you through the process. I recommend *NOSIntro,* by Ian Wade, G3NRW. It's available from the ARRL (see the *ARRL Publications Catalog* in this issue).

You can also get special *NOS* software packages that are designed to "configure" themselves automatically. The programs ask you to enter several items of information, and then proceed to set themselves up in your computer. One such package, *MFNOS* with "Autogen", is available from the Connecticut Digital Radio Association. The software is written for IBM PCs and compatibles.

If you have full Internet access, you can ftp (transfer) *MFNOS* with "Autogen" to your computer by connecting to

ftp.radnet.com. Log in as "anonymous" and enter your Internet address when you're asked for a password.

You'll find the following files in the mfnos directory:

mfnos!ct.zip (for Connecticut stations only)
mfnos!44.zip (for all other stations)
!readme.1st (installation directions)

If you have questions about *MFNOS*, send e-mail to: n1nwp@a3bee2.radnet.com.

If you don't have Internet access, you can send a blank high-density diskette along with a self-addressed, stamped disk envelope to:

Bill Lyman Jr, N1NWP
219 S Orchard St
Wallingford, CT 06492

I've also placed *mfnos!44.zip* and *mfnos!ct.zip* on the ARRL BBS. If you have a modem, call 203-666-0578 and download it.

Once you have the software running in your computer, all you need is your own IP address. These are assigned by IP *address coordinators* throughout the country. You'll find a list of coordinators in ARRL books such as the *Operating Manual* and *Your Packet Companion*. The list can also be found on ham-radio-oriented computer bulletin boards (including the ARRL BBS) and several on-line services.

If you don't want to jump into TCP/IP just yet, take heart. Some gateways offer access via AX.25 packet. You can learn about the services each gateway offers by reading through the gateway "resource" file. See the sidebar, "Finding Your Nearest Gateway."

Using the Gateways

The types of activity you can enjoy through a gateway depend on the gateway operator. Some gateways allow access only by specific amateur stations (such as PBBSs). Others limit individuals to sending and receiving mail, or joining real-time "conferences." Here's a quick rundown of what you might encounter:

❑ ***Electronic Mail***: Some gateways allow you to pass packet mail using the Internet as a high-speed relay. You can send TCP/IP e-mail (known as *SMTP*) from one network to another. You may even be able to send AX.25 packet mail through the gateways. Ask the operator to tell you how to create addresses for e-mail traveling via his or her gateway. The procedures may differ depending on whether you're using TCP/IP or AX.25 packet.

Finding Your Nearest Gateway

When this article went to press, there were more than 30 Amateur Radio/Internet gateway systems in the United States. The trick is finding them. The information *does* exist, although most amateurs are unaware of it. That's about to change.

The list of gateways—including the services they offer, packet frequencies, call signs and so on—can be found in a text file called **resource**. If you have full Internet access, you can transfer (ftp) the file into your computer and read it. Connect to:

ftp.radnet.com

Look in the **gateways** directory.

We've also placed this file on the ARRL BBS (203-666-0578). It's called RESOURCE.TXT. You can obtain the file from the ARRL *InfoServer,* too. You can tap the *InfoServer* by using e-mail from your Internet account, or from any of the on-line services such as America On-Line, CompuServe, and so on. See "Exploring the Internet—*Part 2*" in the October 1994 *QST* for instructions on using the ARRL *InfoServer*.

Once you have the file, read through it and find the gateway closest to you. If you're lucky, you'll find a gateway in the city where you live. If not, you'll need to locate a gateway that you can reach via your local or regional packet network (either on TCP/IP or standard packet). Be careful about trying to reach distant gateways over packet networks. Even on TCP/IP, the throughput gets worse the farther you get from home. You might be able to establish a link, but it will be too slow to be usable.—*WB8IMY*

Gateway operators sometimes shy away from offering e-mail capability because of possible legal complications (violations of third-party traffic rules and so on). Others solve this problem by holding and reviewing all e-mail messages that arrive from the Internet before passing them onto the packet network.

❑ ***FTP***: This function is available to TCP/IP packeteers, but few gateways offer it. FTP stands for *file transfer protocol*. It's a means to move files efficiently from station to station. TCP/IP packeteers use it to send computer programs, images, you name it!

One problem with FTP activity via the gateways is the slow data rates on the radio sides of the links. Even at 9600 bits/s, transferring a large file could take a long time and seriously bog down the gateways.

❑ ***Telnet***: This is another TCP/IP-only function. By using telnet you can access an amateur TCP/IP station remotely through an Internet gateway. That's what I was doing in the example shown at the beginning of this article.

If you're skeptical, try to reach my home station. Establish a telnet link to your nearest gateway station (if it offers telnet capability), then ask it to telnet to **44.88.4.35**. A "host unknown" message means that the system was unable to find a route to me. If this happens, telnet to **N1NWP-1.AMPR.ORG**, then telnet to **WB8IMY**. "Failure with..." usually means that my station was not on the air. Try again later.

❑ ***QSO Bridge***: This is a live, keyboard-to-keyboard roundtable that's on 24 hours a day. If your gateway offers access to the bridge, you can enjoy conversations with hams throughout the world. Most of them are sitting at the keyboards of packet stations just like yours. Thanks to the Internet and the gateway stations, everyone can get together and enjoy the conversational equivalent of a food fight! Figure 2 shows what the QSO bridge looks like as seen through the N1NWP gateway.

The Future of Gateways

You can expect to see more gateways appearing on the air in the near future. The cost of private Internet access is falling, so gateway stations will become more affordable. Gateways are especially attractive to Technicians like N1NWP who have an interest in computers and want to expand their communication horizons.

As the gateways proliferate, more networks will link in this fashion. The benefit to hams will be much faster movement of traffic between regional networks—and even *within* networks where Internet wormholes can fill troublesome gaps.

Some hams may grouse at the idea of using a nonamateur network as part of an Amateur Radio communication system. Until a radio-based alternative is found, however, the Internet stands as a valuable resource for expanding Amateur Radio packet activity. Remote packet stations at disaster sites, for example, could link to an Internet gateway and pass traffic to distant points with extraordinary efficiency.

If you can reach an Internet gateway, explore it and see what it has to offer. Don't forget to send a "thank you" note to the gateway operator when you get a chance. His or her effort—and often *money*—makes your enjoyment possible!

Many thanks to Bill Lyman, N1NWP, for his contribution to this article.

Packet Perspective

Conducted By Stan Horzepa, WA1LOU
One Glen Ave, Wolcott, CT 06716-1442
Internet: horzepa@gdc.com CompuServe: 70645,247

Messages via Packet/Internet Gateways

There are misconceptions and misinformation floating around packet radioland about how to use the gateways that connect amateur packet radio and the Internet. Because no two gateways are exactly alike, it's not possible to provide specific instructions that explain how to use them all. On the other hand, instructions that are too general will be so lacking in information that the reader will get little or nothing out of them.

The third alternative (the one I've chosen to use here) is to provide specific instructions on how to use a particular packet-to-Internet gateway. After reading them, you should understand how to use one gateway and be able to apply that knowledge to other gateways you may want to use. Although other gateways you encounter may not work exactly the same as the one described here, they'll be similar enough so that you won't have to be completely re-educated when you use a different gateway.

Jim Durham, W2XO, Gibsonia, Pennsylvania, runs a popular packet-to-Internet gateway and has generously agreed to allow me to use his gateway as the example for properly using packet-to-Internet gateways. Don't worry; Jim wrote these instructions, so you can be assured that they work (and weren't mangled by me). Here goes:

Using the W2XO Gateway

The gateway is configured to make mailing from the ham packet radio network to and from the Internet as straightforward as possible and in accordance with standard mail gateway practices. You use the subject lines and message text as usual (you don't need any special lines in the text or subject of the message). However, Internet users have to register with me so that I can assign an alias, as explained below.

Mailing from the Internet to Packet Radio

On the Internet, the conventional method of mailing to someone on another network is to replace the "at" symbol (@) in the non-Internet address with a percentage sign (%) and follow the modified non-Internet address with an @ and the gateway host's address. The non-Internet address johndoe@whatnet would be entered on an Internet mailer as johndoe%whatnet@*hostname.domain* where *hostname.domain* is the gateway host's address.

To send a message to packet radio from the Internet, all you have to do is replace the ham packet radio address' @ with a percentage sign, follow that with an @ and then the gateway's Internet address (in this case, w2xo.pgh.pa.us). To mail a message from the Internet to the packet radio address of *WA1YUA@W1EDH.CT.USA.NOAM*, you'd replace its @ with %, follow it with @, and then the gateway's Internet address. *WA1YUA%W1EDH.CT.USA.NOAM@w2xo.pgh.pa.us* is the result.

(My packet BBS is W2XO.#SWPA.PA.USA.NOAM and my Internet host name is w2xo.pgh.pa.us. These are, as the man says, "similar but different." Don't confuse them! One is a packet BBS hierarchical address; the other is an Internet address. Because I'm part of the "us" domain on Internet and my computer's host name is w2xo, they look similar.)

That's the Internet to packet side; now for the other side.

Mailing from Packet Radio to the Internet

The amateur packet radio message format allows only six characters in the recipient's address. I use an "alias" to get around this limitation that prevents you from using an Internet address like f.williamson@foobley.com directly. If the Internet recipient is a ham, I use the recipient's call sign. If the recipient isn't a ham, I create a six-character alias in the form 3PTYXX ("third-party").

To send a message to an Internet recipient from the amateur packet radio network, you'd address the message to *call sign* or *alias @W2XO.#SWPA.PA.USA.NOAM* and the message will automatically be forwarded to the Internet address corresponding to that call sign or alias.

For example, suppose you want to send a message from amateur packet radio to joe@somecomputer.com on the Internet. If Joe is a ham whose call sign is WA1YUA, you'd send your message to Joe using the address WA1YUA@W2XO.#SWPA.PA.USA.NOAM and the gateway will look up WA1YUA in its files and mail the message to the Internet address joe@somecomputer.com. If Joe isn't a ham, I assign him an alias like 3PTYXX, and you send your message to Joe using the address 3PTYXX@W2XO.#SWPA.PA.USA.NOAM, and the gateway will look up 3PTYXX in its files and mail the message to the Internet address joe@somecomputer.com.

Many amateurs believe that a message that originates on packet radio (or travels any part of the way on packet) in the US, and is then relayed overseas via the Internet, falls into the category of *international third-party traffic*, meaning it can only be handled with countries where third-party agreements are in effect—but *this is not true*. According to the Regulatory Information Branch at ARRL HQ, as long as the radio traffic is internal to the US or exchanged with a country with which we have a third-party agreement, it is completely legal. That it came from or is destined for a country with which we do not have an agreement is immaterial, as long as it travels to or from that country via means other than Amateur Radio (eg, the Internet, postal mail, telephone, etc).

Third-party traffic is defined by the FCC as communication between the control operator of one station and the control op of another station on behalf of a third party, so the rules deal with the locations of the control operators involved, not of the third party. Regarding international radio communication, the reason for the rule is to prevent loss of revenue to countries where telecommunications is a state monopoly. So if the message travels via the Internet in those countries, their telecommunication systems have not been bypassed.—*WA1LOU*

Seeking Super Packet Tricks

Last month's column was devoted to super packet tricks. There was a lot of interest in the tricks I discussed and I'd like to bring you more in the future, so send me the super packet tricks you use to get more out of packet radio (or just to get something out of packet radio). I'll present them here and you'll become rich and famous. Well, maybe not rich, but how about becoming famous by enriching others?

KISSes, POPs and Pings

If you're looking for a new challenge, try unraveling the mysteries of TCP/IP.

By Steve Ford, WB8IMY
Assistant Managing Editor
Internet: sford@arrl.org
AMPRNET: 44.88.4.35

re you getting weary of the same old packet? Connect to the local bulletin board (BBS), grab your waiting mail, read a few bulletins, and log off. Now *that's* hamming (yawn)!

Hey, I have nothing against this type of packeteering. I check my local BBS almost every day and read the bulletins, too. But after a while (about a decade, in my case) you get a thirst to do something more.

If you want to expand your packet horizons, try *Transmission Control Protocol/Internet Protocol*—otherwise known as *TCP/IP*. Assuming that you already own a packet station (FM transceiver, computer and terminal node controller [TNC]), the additional cost to run TCP/IP is...*nothing*. Nada. Zip. Nil.

The Internet and Amateur TCP/IP

If you've been fortunate enough to surf the Internet, you've already dabbled in TCP/IP. (Although if you're using a *shell* program that makes the Internet more user friendly, many of the nuts and bolts of TCP/IP are transparent to you.) As complicated as it may sound, TCP/IP is simply a group of packet *protocols* that make it possible for the Internet to shuffle information throughout the world.

But what's a protocol? A protocol is a standardized way of doing something that many parties—people or computers—agree to recognize. For example, your ham club may use *Robert's Rules of Order* to maintain sanity during meetings. (Making "motions" to vote and so on.) Well, those rules are a type of protocol. When you hear Internet users say they're going to *ftp* a file, you're hearing a TCP/IP term. FTP stands for *File Transfer Protocol*. You'll also hear references to SMTP (*Simple Mail Transfer Protocol*) Telnet and others. All these protocols reside under the grand umbrella we call TCP/IP.

TCP/IP was pretty much confined to the Internet and Unix-type computer systems until Phil Karn, KA9Q, decided to adapt it for use on ham networks and IBM PCs. The operating system he created was originally called *NET* In time he rewrote the software and it debuted to the ham populace as *NOSNET*, or just *NOS* for short (*Net Operating System*). Not long thereafter, *NOS* was adapted to run on Macintoshes and other computers.

Of course, hams can't resist the urge to tinker. Talented programmers followed Phil's lead and created their own versions of *NOS*. That's why you'll find so many available. There's *GRINOS*, *JNOS*, *TNOS*, *MFNOS* and so on. All of these *NOS* versions use the same TCP/IP protocols. Even the original Unix jargon remains.

But why go to so much trouble? After you get the software running and the hardware perking along, what do you have?

Simply this: an Amateur Radio version of the Internet.

ftp wb8isz

When the link is established, my terminal displays:

SYN sent
Established
220 wb8isz.ampr.org FTP version 890421.1e ready at Sat Aug 13
18:22:26 19

Now I'm ready to log in. At the command prompt I send: **user anonymous**. My terminal displays WB8ISZ's response.

331 Enter PASS command

No problem. Most systems allow you to use your call sign as the password.

pass wb8imy

When I see **230 Logged in**, it's time to check his computer to see what he has to offer. All I have to do is send **dir** and my screen displays:

200 Port command okay 150 Opening data connection for LIST\public

switch.map	**1,500**	**19:57 02/19/95**
tcp/ip.doc	**10,000**	**02:30 04/01/95**
space.exe	**20,000**	**22:25 04/16/95**

3 files 13,617,152 bytes free. Disk size 33,400,832 bytes
Get complete, 200 bytes received

It looks complicated at first glance, but all it's telling you is that WB8ISZ has three files available for transfer. The name of each file is shown along with its size and the date it was placed on his disk. If you have some experience with computers, this may look familiar. (You've just issued the "directory" command that's common to most machines.)

I happen to know that "space.exe" is a game that WB8ISZ has written for my particular computer. Regardless of whether it's an ASCII or binary file, I can transfer a copy by simply using the **get** command.

get space.exe

Figure 1—A typical ftp (file transfer protocol) session. The characters that actually appear on my computer monitor are shown in bold type.

The Attraction of TCP/IP

When a group of packet stations decide to adopt TCP/IP as a way of communicating among themselves, the resulting network looks a lot like the Internet. We call it the *AMPRNET—Amateur Packet Radio Network*. It operates in much the same fashion as the Internet and, as I've already mentioned, it uses the same protocols. The primary difference is that amateur networks are usually much slower and more limited in scope.

Some TCP/IP networks are very small—just a loose group of dedicated enthusiasts who enjoy swapping information. At the opposite end of the scale, there are vast networks that offer coverage to packeteers in several states. They achieve this coverage through the use of special TCP/IP nodes (often called *switches*) that act as relays.

Several of these large networks include *gateways* to transfer information to and from the Internet. The same gateways can also function as *wormholes*—Internet pipelines that link distant sections of large networks.

But what makes TCP/IP so special? Here are just a few of the high points...

Mail—TCP/IP mail is sent from station to station through the network. There are no BBSs involved. You need only prepare a message and leave it in your own TCP/IP "mailbox." Within seconds your computer will attempt to make a connection to the target station and deliver the message. If the target station isn't on the air, your computer holds the message and keeps trying. When you check your computer and see that the message is no longer in the mailbox, you can be sure it has arrived safely at the other station.

File Transfers—You can use the file-transfer protocol (FTP) to easily transfer files (games, images, or whatever) to any station on the network (see Figure 1).

Data Handling—Rather than spewing data at random intervals like standard packet, TCP/IP stations take the smart approach and automatically *adapt* to network delays. As the network slows down, TCP/IP stations sense the change and lengthen their transmission delays accordingly. (They don't transmit as often.) As the network speeds up, the TCP/IP stations shorten their delays to match the pace (they transmit more frequently). This kind of intelligent network sharing virtually guarantees that all packets will reach their destinations with the greatest efficiency the network can provide.

Direct Addressing—Every TCP/IP station in your network has an address. These *IP* addresses are assigned by volunteer coordinators throughout the country (see the sidebar, "Your Own IP Address").

My address, for example, is 44.88.4.35. Reading from left to right, 44 tells you that this is the address of an Amateur Radio TCP/IP station (rather than a nonham Internet site). The 88 designates my part of the New England regional network (Connecticut). The 4 gets even more specific, pointing to my little corner of the state (known as a *subnet*). Finally, the 35 is my unique address in the subnet.

But you don't need to know the addresses of your fellow TCP/IP users! All of this information is contained within your *NOS* software in a file called *DOMAIN.TXT*. When I want to send a message to, say, WS1O, I simply address the message to WS1O. When it's time to send the message, *NOS* will search through DOMAIN.TXT and locate WS1O. When it does, it'll use the IP address it finds there.

By analyzing the IP address, the switches and other stations probe through the network to create a link to WS1O. Once the "circuit" is established, the data flows and the message is delivered. All of this takes place without your lifting a finger!

Many versions of *NOS* contain DOMAIN.TXT. It may be a small file that includes only the users in a particular area. On the other hand, it could be the granddaddy of all DOMAIN.TXTs that lists every amateur TCP/IP user in the world! If your *NOS* doesn't have a DOMAIN file, you'll need to find a copy. See the sidebar, "Getting Started in Four Steps."

Multitasking—With TCP/IP you can do several things simultaneously. For example, you can send a message, receive a message and transfer a file—all at the same time!

Flexibility—*NOS* provides an excellent platform for developing new protocols that will dramatically expand the capabilities of the AMPRNET. Some versions already support Gopher and an amateur version of the World Wide Web is on the way.

Ping! Is Anyone Home?

Is your buddy on the air this afternoon? Should you bother trying to get a message

Your Own IP Address

Before you can get on the air with TCP/IP, you need your own *AMPRNET* IP address. These addresses are assigned by the volunteer coordinators listed below. This list is subject to change without notice. If you have Internet ftp capability, you'll find updated *AMPRNET* coordinator lists at **ftp.ucsd.edu**.

AK	John Stannard, KL7JL
AL	Richard Elling, KB4HB
AR	Richard Duncan, WD5B
AZ	David Dodell, WB7TPY
CA, Antelope Valley/Kern County	Dana Myers, KK6JQ
CA, Los Angeles—San Fernando Valley	Jeff Angus, WA6FWI
CA, Orange County	Terry Neal, AA6TN
CA, Sacramento	Bob Meyer, K6RTV
CA, Santa Barbara/Ventura	Don Jacob, WB5EKU
CA, San Bernardino and Riverside	Geoffrey Joy, KE6QH
CA, San Diego	Brian Kantor, WB6CYT
CA, Silicon Valley—San Francisco	Douglas Thom, N6OYU
CO (north)	Fred Schneider, KØYUM
CO (south)	Bdale Garbee, N3EUA
CO (west)	Bob Ludtke, K9MWM
CT	Bill Lyman, N1NWP
DC	Richard Cramer, N4YDP
DE	Butch Rollins, NF3F
FL	Brian Lantz, KO4KS
GA	Doug Reed, N3AIA
HI and Pacific islands	John Shalamskas, KJ9U
ID, eastern WA	Steven King, KD7RO
IL (central and southern)	Chuck Henderson, WB9UUS
IL (northern)	Ken Stritzel, WA9AEK
KS	Dale Puckett, KØHYD
KY	Allan Dayton, NØKFO
LA	James Dugal, N5KNX
MA (center and eastern)	Gordon LaPoint, N1MGO
MA (western)	Bob Wilson, KA1XN
MD	Howard Leadmon, WB3FFV
ME	Carl Ingerson, N1DXM
MI (upper peninsula)	Thomas Landmann, N9UDL
MI (lower peninsula)	Jeff King, WB8WKA
MN	Andy Warner, NØREN
MO	Stan Wilson, AKØB
MS	John Martin, KB5GGO
MT	Steven Elwood, N7GXP
NC (eastern)	Mark Bitterlich, WA3JPY
NC (western)	Charles Layno, WB4WOR
ND	Steven Elwood, N7GXP
NE	Mike Nickolaus, NFØN
NH	Gary Grebus, K8LT
NJ (northern)	Dave Trulli, NN2Z
NJ (southern)	Bob Applegate, WA2ZZX
NM	J. Gary Bender, WS5N
NY (eastern)	Bob Bellini, N2IGU
NY (western)	Dave Brown, N2RJT
New York City and Long Island	Bob Foxworth, K2EUH
NV (southern)	Earl Petersen, KF7TI
NV (northern)	Bill Healy, N8KHN
OH	John Ackerman, AG9V
OK	Joe Buswell, K5JB
OR	Ron Henderson, WA7TAS
OR (northwest and Vancouver, WA)	Tom Kloos, WS7S
PA (eastern)	Doug Crompton, WA3DSP
PA (western)	Bob Hoffman, N3CVL
PR	Karl Wagner, KP4QG
RI	Charles Greene, W1CG
SC	Mike Abbott, N4QXV
SD	Steven Elwood, N7GXP
TN	Jeff Austin, K9JA
TX (northern)	Jack Snodgrass, KF5MG
TX (southern)	Kurt Freiberger, WB5BBW
TX (western)	Rod Huckabay, KA5EJX
UT	Matt Simmons, KG7MH
VA	Jim DeArras, WA4ONG
VA (Charlottesville)	Jon Gefaell, KD4CQY
VI	Bernie McDonnell, NP2W
VT	Ralph Stetson, KD1R
WA (eastern)	Steven King, KD7RO
WA (western)	Bob Donnell, KD7NM
WI	Thomas Landmann, N9UDL
WV	Rich Clemens, KB8AOB
WY	Reid Fletcher, WB7CJO

to him, or transfer a file to his computer? TCP/IP gives you an easy way to find out. Using WS1O as our example, I can determine if his station is active by issuing a *ping*.

ping ws1o

My station sends a response request to WS1O. It weaves through the network until it arrives at his station. His computer responds and I receive a reply, along with the time (in milliseconds) that it took for my ping to get there and back.

44.88.4.23: echo reply 7000 ms

All this jargon basically tells me that WS1O is on the air (his address is shown) and that my ping required 7000 milliseconds (7 seconds) to make the trip.

Give Him a Finger

No, not *that* finger! Let's say that you want to find out more about WB8IMY without bothering to send a message and wait for my reply. The quick way to do this is through the *finger* command. It's a terrific feature for nosy packeteers!

finger wb8imy@wb8imy

Assuming that you have access to my TCP/IP network, your finger request will travel to my station. Like many TCP/IP operators, I have a special text file that provides a brief rundown on my station. When my computer receives a finger request, this file is sent automatically.

Connect with Telnet

The *telnet* function allows you to do keyboard-to-keyboard work. You can connect to another station on the network and access his or her mailbox, get a list of stations heard and so on. If the other operator is present at the keyboard, you can even enjoy a "live" conversation.

Many TCP/IP Internet gateways allow you to use telnet to connect to an Internet Relay Chat (*IRC*). This is the equivalent of a conversational food fight in cyberspace! (Everyone is talking to each other at once.) If you don't mind the confusion, you'll enjoy tapping into an IRC.

TCP/IP Q&A

Q: Can I use *any* TNC with TCP/IP?

A: As long as it includes the KISS mode, yes. (Most TNCs do.) To put your TNC in KISS, you may have to enter the command **KISS ON** followed by **RESET**. To get out of KISS, some TNCs require you to hold the **ALT** key and type 192. Hold **ALT** again and type 255. Release the **ALT** key and then press it one more time while typing 192. These procedures vary, so consult your TNC manual.

Q: Does it have to be a 9600-bit/s TNC?

A: No. You'll find plenty of 1200-bit/s TCP/IP activity, primarily on 2 meters. If you can upgrade to a 9600-bit/s station, so much the better. TCP/IP really shines at high data rates! Most 9600-bit/s TCP/IP activity seems to take place on the 440-MHz band.

Q: What kind of computer do I need?

A: I recommend an AT-class IBM PC or compatible such as a 286, 386 or 486. Some versions of *NOS* will run on old XTs, but others won't. Some TCP/IP operators also enjoy using Apple Macintosh machines.

Q: Do I have to leave my computer on 24 hours a day to receive mail?

A: If you can leave your TCP/IP station running continuously, you'll be able to send and receive mail at any time. This is convenient for everyone concerned. Some packeteers buy used PCs (such as relatively cheap 286 machines) and dedicate them to their TCP/IP stations. They just park them in a corner with the other equipment and leave them on day and night.

This approach isn't practical for everyone, though. That's where *POP* comes into play. POP stands for *Post Office Protocol*. If another station on your network is active 24 hours a day, he or she may be willing to act as a depository for your incoming mail—otherwise known as a POP *server*. By activating the POP function in your *NOS* software, your station will automatically contact your POP server and grab any waiting mail. This usually takes place immediately after your station comes on the air.

Q: Can I still connect to my friends on standard (AX.25) packet if I'm running *NOS*?

A: Absolutely! You can still connect to your AX.25 BBS and your friends can connect to you. For example, the version of *NOS* I use allows me to make an AX.25 connect to my local BBS by typing the following command:

c 2m w1nrg

The letter "c" stands for "connect" while "2m" is the designator my *NOS* uses to access my TNC. Finally, "w1nrg" is the call sign of my local packet BBS. When I enter this command, *NOS* will use my TNC to make a standard packet connection to W1NRG.

Some AX.25 packet BBSs even have "ports" to the TCP/IP network. This means that you can use your TCP/IP telnet command to access the BBS without resorting to the command line shown above, and without leaving your TCP/IP network frequency. It's as though the BBS is a split personality—it's a TCP/IP system *and* a standard packet BBS!

Special thanks to John Ackerman, AG9V, for his assistance in the preparation of this article.

Getting Started in Four Steps

1. If you don't own a packet station now, you'll need to build one. All that's required is a computer, a 2-meter FM transceiver and a *terminal node controller*, or *TNC*. The TNC must be capable of operating in the KISS mode (*K*eep *I*t *S*imple, *S*tupid—no kidding!). All TNCs made within the last eight years or more include KISS. Beware that some TNC emulation systems such as *DIGICOM*, *Poor Man's Packet* and *BAYCOM* may not be compatible with TCP/IP software.

2. Get a version of *NOS*. If you can determine the version that's most popular on your network, it's a good idea to use the same. It's much easier for the locals to help you if you're using similar software (see Step 4).

You'll find *NOS* software on many on-line services such as CompuServe (in the *HamNET* forum) and on a number of ham-oriented computer bulletin boards. (Yes, the ARRL BBS has several *NOS* versions available. Call 203-666-0578 and download one of them!) If you have Internet capability, you can ftp many versions of *NOS* from **oak.oakland.edu** or **ftp.ucsd.edu**. *NOS* is also offered for a nominal price by TAPR (Tucson Amateur Packet Radio). For a complete list, send a self-addressed stamped envelope to: TAPR, 8987-309 Tanque Verde Rd, No. 337, Tucson, AZ 85749-9399.

3. Get an IP address. Contact one of the coordinators shown. Some coordinators may prefer to issue a temporary "test" address at first. If you find that you like TCP/IP, they may ask that you send a message to them via the network and request a permanent address. They're not making you jump through hoops for their enjoyment. It's just that they don't want to assign permanent addresses to hams who are not going to be active on the network for the long term.

4. Install your *NOS* and get on the air. I *strongly* recommend that you get outside help from an experienced TCP/IP operator for this step. You must configure your AUTOEXEC.NOS file with your IP address, routing parameters (who will relay your packets?) and several other bits of information. This can be a frustrating, time-consuming exercise if you don't know what you're doing!

Local TCP/IP operators are your best resource, by far. They'll be able to help you get your station on the air in the shortest time. They'll also have the most up-to-date version of DOMAIN.TXT for your particular network. TCP/IP packeteers are almost evangelical in their quest for new converts, so you'll likely find several willing to assist. If you're already active on standard packet, drop a message on your local bulletin board and ask for assistance. TCP/IPers usually check the local BBSs regularly.

If you must install *NOS* by yourself, pick up a copy of *NOSintro* by Ian Wade, G3NRW. This book will supply much of the information you need. See your favorite dealer or the *ARRL Publications Catalog* in this issue.

Exploring the 9600-Baud PACSATs

Look to the heavens for a new packet radio challenge.

By Andrew Cornwall, VE1COR
5 Belmore Dr
Wellington, NS B2T 1J4
Canada

Our Amateur Radio packet satellite "fleet" is impressive. There are presently eight PACSATs in orbit serving thousands of hams worldwide. Five of these satellites operate at data rates of 1200 bit/s. (Most of the 1200-bits/s satellites require special *PSK* TNCs in ground stations, not ordinary packet TNCs.) The remaining three—OSCARs 22, 23 and 25—use 9600 bit/s.

The 9600-bit/s PACSATs are the most popular. Why? All of the PACSATs travel in relatively low orbits. This means that you can use them for only about 15 or 20 minutes before they disappear below your horizon. If you only have a short amount of time on the satellite, you want to communicate at the highest data rate you can manage. You can accomplish eight times as much in 15 minutes at 9600 bit/s as you can at 1200 bit/s!

Profile of the Big Three

UoSAT-OSCAR 22 was designed and built by the University of Surrey (England) in 1991. KITSAT-OSCAR 23 and KITSAT-OSCAR 25 were designed by the Korean Advanced Institute of Science and Technology (KAIST) with assistance from the University of Surrey. They were launched in 1992 and 1993, respectively.

All three satellites have store-and-forward messaging as their primary function. That is, they accept messages and hold them until they are downloaded by the recipients. They also take still pictures of the Earth for downloading as computer graphics, and run telemetry for scientific experiments. UO-22, KO-23, and KO-25 use the PACSAT protocol and operate in Mode J, which means they receive on 2 meters and transmit on 70 cm. FM is used for both the uplinks and downlinks.

One of the fascinating aspects of these satellites is their small size. They pack all of the equipment for radios, power, computer, image capturing, thermal control, and physical stabilization in a 9-inch, solar-cell-clad cube. Two-meter and 70-cm antennas are attached at the top and bottom of the cube. The UHF transmitter output is about 5 W.

Each satellite orbits the Earth in a path that crosses both poles at an altitude of somewhat less than 800 km for UO-22 and KO-25, and about 1,300 km for KO-23. As they pass overhead, ground stations within their 5,000-km-wide *footprints* have an opportunity to upload and download packet data files consisting of personal messages, bulletins, computer programs and graphics. The files range in size from a few hundred bytes to over 300 kbytes. Even at 9600 bit/s you may need a few passes to download a large file.

"Broadcast" Protocol

Despite the significant amount of data that can be captured during a pass, there is considerable competition among ground stations about exactly *which* data the satellite should receive or send! There are typically two or three dozen stations within a satellite's moving footprint, all making their requests.

The PACSATs sort out the pileups by creating two *queues* (waiting lines)—one for uploading and another for downloading. The upload queue can accommodate two stations and the download queue can take as many as 20. Once the satellite adds a ground station to the queue for downloading, the ground station moves forward in the line until it reaches the front, whereupon the satellite services the request for several seconds.

Unless the file you want is small, you won't get it all in one shot. After the satellite completes its transmission to your station, you're bounced back to the end of the queue once again. In some cases you won't automatically be put in queue again, but will just fall out of line entirely. When this happens, your station must signal the satellite, sometimes repeatedly, to get back to the end of the queue. For uploading, the satellite

```
MSPE [KO-23]
File  Setup  Directory  Fill  Satellite  Send Msg                Help
12C7A  85   0      0   File 12ECB downloaded
12DE9  29   0      0   File 12E9F heard
12EBA  10  37  85408   Saving file 12E87
12E28   1   0      0   Auto: Request fill of file 12E28
12CE0   2   0      0   NO -1 VE1COR
12E8A   9   0      0   Broadcast queue full.
12E9F   9   0      0   File 12CE0 heard
12D03  16   0      0   File 12C7A heard
12CF4   5  20 106433   Auto: Request fill of file 12E28
12D53  44   0      0   File 12CE0 heard
                       File 12EBA heard
                       Auto: Request fill of file 12E28
Sat Apr 01 22:42:23 1995 Up: 54/21:7 EDAC= 717 F:78960 L:77648 d:0 s:0 b:2072052
Sat Apr 01 22:42:38 1995 Up: 54/21:8 EDAC= 717 F:78192 L:77648 d:0 s:0 b:2072061
Sat Apr 01 22:42:53 1995 Up: 54/21:8 EDAC= 717 F:78192 L:77648 d:0 s:0 b:2072069
Sat Apr 01 22:43:08 1995 Up: 54/21:8 EDAC= 717 F:78192 L:77648 d:0 s:0 b:2072077
Sat Apr 01 22:43:23 1995 Up: 54/21:8 EDAC= 717 F:78192 L:77648 d:0 s:0 b:2072088
Sat Apr 01 22:43:38 1995 Up: 54/21:9 EDAC= 717 F:77488 L:77488 d:0 s:0 b:2072099
Sat Apr 01 22:43:53 1995 Up: 54/21:9 EDAC= 717 F:77488 L:77488 d:0 s:0 b:2072109
Sat Apr 01 22:44:08 1995 Up: 54/21:9 EDAC= 717 F:77488 L:77488 d:0 s:0 b:2072121
Sat Apr 01 22:44:23 1995 Up: 54/21:9 EDAC= 717 F:77488 L:77488 d:0 s:0 b:2072130
Sat Apr 01 22:44:38 1995 Up: 54/21:10 EDAC= 717 F:77488 L:77488 d:0 s:0 b:207214
Sat Apr 01 22:44:53 1995 Up: 54/21:10 EDAC= 717 F:77488 L:77488 d:0 s:0 b:207215
Sat Apr 01 22:45:08 1995 Up: 54/21:10 EDAC= 717 F:78240 L:77488 d:0 s:0 b:207215
Sat Apr 01 22:45:23 1995 Up: 54/21:10 EDAC= 717 F:78240 L:77488 d:0 s:0 b:207216
Sat Apr 01 22:45:38 1995 Up: 54/21:11 EDAC= 717 F:77488 L:77488 d:0 s:0 b:207218
Sat Apr 01 22:45:53 1995 Up: 54/21:11 EDAC= 717 F:77488 L:77488 d:0 s:0 b:207218
Sat Apr 01 22:46:23 1995 Up: 54/21:11 EDAC= 717 F:77488 L:77488 d:0 s:0 b:207220
Sat Apr 01 22:46:38 1995 Up: 54/21:12 EDAC= 717 F:77488 L:77488 d:0 s:0 b:207221
Sat Apr 01 22:46:53 1995 Up: 54/21:12 EDAC= 717 F:78240 L:77488 d:0 s:0 b:207222
PB: WB1HBU VE3BDR ON6XY\D K8TL DC8AM K4UPL ON4KVI\D WB2YLR    Open  2a : GU6EFB
WA8EBM WB2REM NZ3F\D DF5DP G0SUL F1TIV G3CDK\D LX1BB VE3EGO
G3ZUM\D VE3FRH
DIR 44 holes | AUTO 12E28 | 83% | T:540631 | D:40934 | F:375225
```

Figure 1—A lot is going on in this sample of a *WISP* screen during an actual KO-23 pass. The list near the bottom shows the ground stations waiting for downloads of directories (\D), messages, or files. GU6EFB, bottom right, is uploading a file.

Figure 2—Tilting the log periodic antenna to take advantage of its wide vertical beamwidth avoids the need for an expensive azimuth/elevation rotor. Despite the odd angle, the antenna also provides good terrestrial VHF and UHF performance.

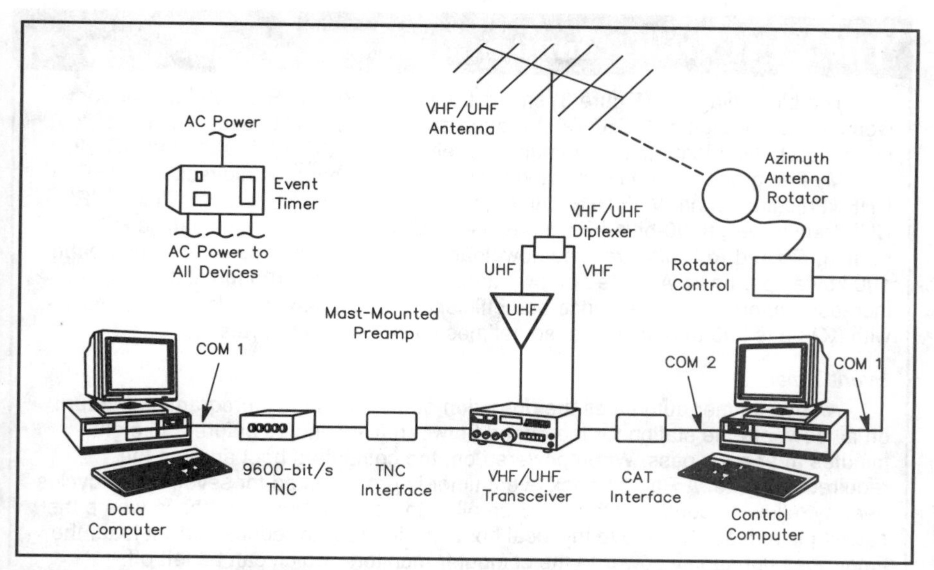

Figure 3—My unusual PACSAT station setup uses two computers. With second-hand PCs, it's also cost-effective. The alternative is to use one properly equipped 386 or 486 PC running *WISP* in Microsoft *Windows* to control the equipment *and* capture the data. Most hams prefer the single-computer approach. I just happened to have two computers available!

attempts to receive from the two ground stations in the upload queue until their various transactions are completed.

If the satellite disappears before you receive the complete file, there's no need to worry. Your PACSAT ground station software "remembers" which parts of the file you still need from the bird. When it appears again, your software can request that these "holes" be filled.

And while all of this going on, *you're receiving data that other stations have requested!* Not only do you get the file you wanted, you also receive a large portion of the data that other hams have requested. In fact, you may receive a number of messages and files without transmitting a single watt of RF. All you have to do is listen. That's why they call it "broadcast" protocol.

Station Equipment and Software

The computer is at the heart of any PACSAT station. The ground station software communicates with the satellite, receiving the broadcast messages, filling holes in files and so on. For years the most popular ground station software was actually two programs: *PB* for downloading and *PG* for uploading. Within the last year or so, however, a new program known as *WISP* has taken the PACSAT world by storm. Written for Microsoft *Windows, WISP* offers a much "friendlier" environment than *PB* or *PG* (see Figure 1). It will also totally automate your station, if you choose to do so. Both *PB/PG* and *WISP* are available from AMSAT, PO Box 27, Washington, DC 20044. AMSAT also sells a valuable guide for anyone who wants to try the PACSATs. It's the *Digital Satellite Guide* and it includes the *PB/PG* software. Send a self-addressed, stamped envelope to AMSAT for more information.

Through software such as *WISP*, and the addition of a special tracking card or external controller, your computer can automate all vital functions. It can run the TNC, tune the radio (and continually *retune* it to track the satellite signal, compensating for Doppler shift), and aim the antenna to track the satellite as it travels across the sky, from horizon to horizon. With a fully automated ground station, there is little for an operator to do. Indeed, many stations run unattended during a pass. In advance of a pass, the human operator of a ground station would schedule the computer to upload and download particular messages and other files. Items received are stored on the ground station's hard disk to be read or otherwise used later.

Ground stations working the 9600-bit/s PACSATs use various kinds of radio and computer equipment. There is not yet a "standard" satellite ground station, and nearly every rig has some aspect of home brew construction.

There are a number of affordable 9600-bit/s packet TNCs on the market. The challenge, however, is to get your radio to transmit and receive the high-speed data signal without distortion. Unlike 1200-bit/s packet, you can't simply run the transmit audio into the microphone jack. Instead, the data signal must be injected at the modulator. The receive signal must be tapped at the FM discriminator. For most kinds of radios the modification is fairly simple. Nevertheless, many ground station operators have experienced the trepidation of exposing the guts of expensive radios, with soldering iron in hand.

Some radios claim to be "9600 ready" without modification, but some work significantly better than others. (See "'9600-Ready Radios': Ready or Not?", by Jon Bloom, KE3Z, in the May 1995 *QST*.) And even the radios that are truly 9600 capable may not be suitable for PACSAT operation. So, many satellite packeteers still rely on rigs they modify themselves, such as the Yaesu FT-736, Kenwood TS-790 or the ICOM IC-820H (only a slight mod is necessary for the IC-820H).

The good news is that you don't need elaborate antennas or high output power. My station is currently set up to work the 9600-baud PACSAT satellites exclusively. I run only 25 W for the 2-meter uplink. My antenna system consists of a single log periodic dipole array with a duplexer to separate the 2-meter and 70-cm signals. Because the elevation beamwidth of the LPDA is quite wide, I've been success-

The VE1COR Ground Station

The block diagram (Figure 3) shows the current VE1COR ground station for working the 9600-bit/s PACSATs. Although my setup is a bit unusual (note the *two* computers), it's adequate to communicate reliably with KO-23, KO-25 and UO-22.

My first ground station consisted simply of a dual-band omnidirectional VHF/UHF vertical antenna, VHF/UHF duplexer, mast-mounted UHF preamplifier, VHF/UHF transceiver, 9600-bit/s FSK modem and TNC, and an IBM-compatible PC running *PB* and *PG* software. The download capability of the this setup was around 100 kbytes per pass. As noted below, additions to the initial ground station have increased operating convenience and efficiency. My present download performance with KO-23 is 300 to 600 kbytes, sometimes more than 700 kbytes.

Event Timer

An event timer automates the operation of the station. It's programmed to turn on ac power to the station for the period between five minutes before and eight minutes after each pass. When powered on, the computers boot and run the required software. A Radio Shack event timer is programmed for seven on/off cycles per day, allowing contact with KO-23 on all of the stronger passes (those where the satellite is at least 15° above my local horizon). In order to reduce current load, the timer does not supply power to the computer monitors, which can be left off.

TNC Interface

The TNC interface is a home brew experiment consisting of several op-amp stages to buffer and match the impedances of the modem and the transceiver. The interface provides a small treble boost to the audio from the received signal. There is also a manual switch that decouples the modem from the transceiver, so the transceiver can be used for terrestrial communications. Most stations would not need a TNC Interface. A single PC is dedicated to talking to the TNC and running the PACSAT communication software.

Computer-Assisted Transceiver (CAT) Control

A commercial CAT interface adjusts for signal voltage differences between the microcomputer's serial port and the transceiver's CAT port. Many functions of the transceiver can be controlled by a microcomputer. In this instance, the computer adjusts the transceiver's receive frequency to offset Doppler shift. The second station computer does this by monitoring the transceiver's deviation meter and changing the receive frequency to keep the meter at center scale. Tracking the received frequency throughout a pass significantly increases download performance.

Log-Periodic Antenna

A log-periodic dipole array offers multiband capability, gain of about 12 dBi, relatively wide beamwidth, and compact size. The added gain improves communication throughput. The antenna is directional, however, and must be aimed at the satellite as it crosses the sky.

Antenna Rotor Control

An inexpensive, light-duty azimuth rotator turns the log-periodic antenna in the horizontal plane. Elevation tracking is not necessary because of the wide vertical beamwidth (60°) of the vertically polarized antenna. Pointing the antenna 22° upward works quite well. During some passes, when the satellite goes directly overhead, the signal drops off for a couple of minutes. With the majority of passes, however, signal interruption is not significant.

The same IBM-compatible computer that tunes the transceiver for Doppler shift also controls the rotator. A home brew rotator controller replaces the factory-supplied manual control box. The microcomputer turns the rotor to predetermined bearings at specific times during a pass. A batch of bearings are computed beforehand by a satellite tracking program and stored in computer files.

ful by tilting the antenna upward at a fixed elevation angle and using only a single rotor for azimuth tracking (see Figure 2). During my early experiments I was able to work UO-22 and KO-23 with just a simple 2-meter/70-cm, omnidirectional vertical antenna! (See the sidebar, "The VE1COR Ground Station.")

A Challenging Hobby!

A common bond among amateurs on satellites generally is the realization that someday their favorite birds will cease to function. This is inevitable, and don't count on anyone to take a ride on the shuttle to fix a broken ham satellite. For example, when KO-23 was launched, its life expectancy was only three to five years because of the expected degradation caused by its orbit being in the Van Allen radiation belt. Countering this pessimistic outlook, however, is the strong interest to launch new satellites. In late March, a Russian rocket failed in an attempt to launch two new PACSATs, one built by the University of Mexico and the other by the Technion-Israel Institute of Technology (see the "Amateur Satellite Communication" column in this issue for details). On the other hand, the Japan Amateur Satellite Organization plans to launch a 9600-bit/s PACSAT called JAS-2 in 1996. At about the same time, AMSAT plans to launch a multipurpose super-satellite known as Phase 3D. Other organizations are also working on Amateur Radio satellite projects. (Did you read the article "Phase 3D: The Ultimate *EasySat*" in the May 1995 New Ham Companion? If not, go back and read it—especially the last few paragraphs!)

Within the inevitable budget constraints, volunteer satellite designers and assemblers attempt to use component parts that are hardened to withstand the massive g-forces and vibration of the launch, the rigors of continual cosmic radiation, and the cyclic extremes of temperature. Each successful pass of a satellite testifies to the ability of the crew that brought it into existence and who continue to maintain system operation.

Data satellites occasionally experience temporary difficulties. KO-23 and KO-25 both "crashed" (in the computer sense) in December 1994. KAIST was successful in reloading the satellites' software and rebooting. KO-23 crashed again after only nine hours, and KO-25 stayed operational for about a week before crashing again. Both satellites were functioning when this article went to press.

Be forewarned that pursuing the amateur satellite hobby, whether voice, CW, or digital, can be inconvenient and frustrating. Communication is possible at only certain, often relatively brief, periods of the day. Working in the digital mode presents the additional challenge of setting up and operating extra gear consisting of microcomputers, software, and the TNC. The complexity of digital ground station operation is especially evident if a problem occurs. For example, when confronted with the problem of poor performance, is it due to the antenna, rotor, antenna preamp, transmission lines, radio, TNC or what? Or is the problem with the satellite itself?

Despite these tribulations, users of UO-22, KO-23, and KO-25 frequently post messages reflecting their enjoyment. To paraphrase one PACSAT devotee in Yorkshire, England: This is no hobby. It is an obsession!

Adapted from an article originally written for the Halifax, Nova Scotia, Amateur Radio Club Newsletter.

By Steve Ford, WB8IMY

DX *PacketClusters*

The time-saving solution for busy DXers and contesters!

I'd love to spend hours trolling the HF bands. I really would. With a hot cup of coffee at my side, I could spend these chilly winter evenings searching for DX on my favorite bands. And then there are the contests. You can burn through several hours hunting for that last state, section, country or whatever.

In truth, however, I'm typical of most middle-aged and younger hams—the majority of the ham population today. Our evenings and weekends are filled with school or family activities. We have to squeeze Amateur Radio into those few precious moments when we can park ourselves in front of our radios.

So how do you reconcile the desperate need to pursue DX or contests, and still find time to meet all your other commitments? Wouldn't it be nice if someone could at least do the hunting for you? That's the most time-consuming part. With the reports of their "reconnaissance" in hand, you could choose which "prey" you wanted to stalk.

But who would be crazy enough to volunteer for such a job? As it turns out, there are thousands of such "volunteers" throughout the nation...and thanks to some nifty software developed by Dick Newell, AK1A, they're connected to *DX PacketClusters*!

As Close as your TNC

If you have a packet radio setup (a 2-meter FM transceiver, terminal node controller [TNC] and a computer or data terminal), you're ready to enjoy the advantages of *PacketClusters*. A *PacketCluster* is a network of relaying nodes devoted to contesting and DX hunting. Some *PacketCluster* networks are small, single-node systems with a dozen users or so (see Figure 1). Other networks are vast, with many nodes interconnected using VHF/UHF or Internet links (see Figure 2).

Regardless of its size, a *PacketCluster* exists for one fundamental purpose: *sharing information*. When a ham sends a piece of information to a *PacketCluster* node (the appearance of a rare DX station, for example), the node relays it to everyone connected to the network. If you're on the network, you share the "profits."

Busy hams like me can connect to their nearest *PacketCluster* nodes and, within seconds, receive lists of all the most recent DX sightings. In the blink of an eye I can see which bands are hot at the moment, and which DX stations are on the air (and at what frequencies). And if I'm the lucky guy who stumbles across a DX signal first, I can post the information for everyone else to see. (Share and share alike!) If I'm dabbling in a contest, I can connect to the *PacketCluster* and determine where to find the stations I need to boost my score.

The Basics

You connect to a *PacketCluster* in the same way you'd connect to any other packet station. However, the information you'll receive will be very different!

The secret of using a *PacketCluster* is knowing the various commands. The most common ones are shown in Table 1.

My routine is to connect to my nearest node and immediately ask for a list of the latest DX sightings using the **SHOW/DX** command (Figure 3). Then, I take a peek at solar conditions by sending **SHOW/WWV**. By sending these two commands, I receive a capsule summary of band conditions and DX activity.

If I'm tuning through the bands I might stumble on a DX station worthy of a *spot* on the cluster. Let's say I hear SV3AQR on

Figure 1—A small *PacketCluster* may be comprised of a single node. Everyone who connects to the node shares the same DX or contest information.

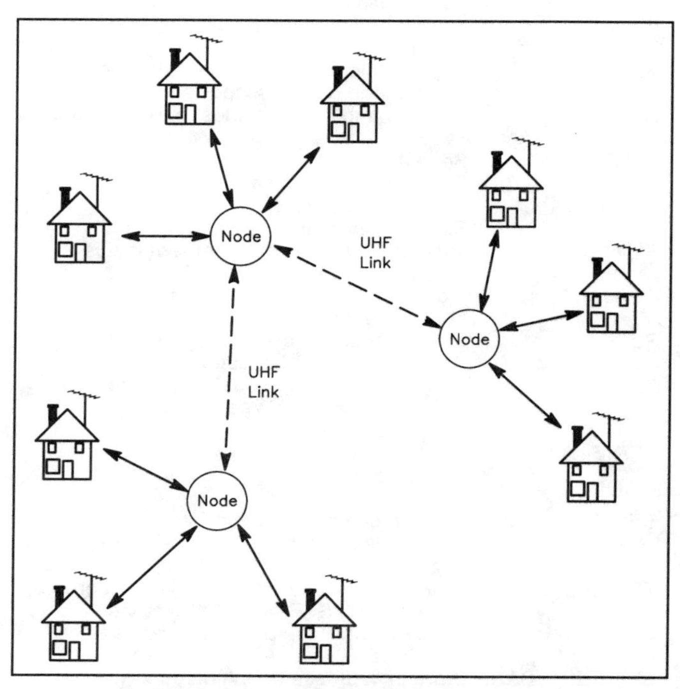

Figure 2—Other *PacketCluster* networks are huge, covering entire states or regions. The individual nodes often use UHF radio links or Internet connections to exchange information.

Table 1

Common *PacketCluster* Commands

Command	Description
ANNOUNCE	Make an announcement.
A *x*	Send message *x* to all stations connected to the local node.
A/F *x*	Send message *x* to all stations connected to the cluster.
A/*x y*	Send message *y* to stations connected to node *x*.
A/*x y*	Send message *y* to stations on distribution list *x*.
BYE	Disconnect from cluster.
B	Disconnect from cluster.
CONFERENCE	Enter the conference mode on the local node.
CONFER	Enter the conference mode on the local node. Send <CTRL-Z> or /EXIT to terminate conference mode.
CONFER/ F	Enter the conference mode on the cluster. Send <CTRL-Z> or /EXIT to terminate conference mode.
DELETE	Delete a message.
DE	Delete last message you read.
DE *n*	Delete message numbered *n*.
DIRECTORY	List active messages on local node.
DIR/ALL	List all active messages on local node.
DIR/BULLETIN	List active messages addressed to "all."
DIR/*n*	List the *n* most recent active messages.
DIR/NEW	List active messages added since you last invoked the DIR command.
DIR/OWN	List active messages addressed from or to you.
DX	Announce DX station.
DX *x y z*	Announce DX station whose call sign is *x* on frequency *y* followed by comment *z*, eg, DX SP1N 14.205 up 2.
DX/*a x y z*	Announce DX station whose call sign is *x* on frequency *y* followed by comment *z* with credit given to station whose call sign is *a*, eg, DX/K1CC SP1N 14.205 up 2.
FINDFILE	Find file.
FI *x*	Ask the node to find file named *x*.
HELP or ?	Display a summary of all commands.
HELP *x*	Display help for command *x*.
READ	Read message.
R	Read oldest message not read by you.
Rn	Read message numbered *n*.
R/*x y*	Read file named *y* stored in file area named *x*.
REPLY	Reply to the last message read by you.
REP	Reply to the last message read by you.
REP/D	Reply to and delete the last message read by you
SEND	Send a message.
S/P	Send a private message.
S/NOP	Send a public message.
SET	Set user-specific parameters.
SE/A	Indicate that your computer/terminal is ANSI-compatible.
SE/A/ALT	Indicate that your computer/terminal is reverse video ANSI-compatible.
SE/H	Indicate that you are in your radio shack.
SE/L *a b c d e f*	Set your station's latitude as: *a* degree *b* minutes *c* north or south and longitude *d* degrees *e* minutes *f* east or west.
SE/N x	Set your name as x.
SE/NEED x	Store in database that you need country(s) whose prefix(s) is *x* on CW and SSB. eg, SE/NEED XX9.
SE/NEED/BAND =(*x*)*y*	Store in database that on frequency band(s) x, you need country(s) whose prefix(s) is y, eg, SE/NEED/BAND=(10)YA.
SE/NEED/*x y*	Store in database that in mode *x* (where *x* equals CW, SSB or RTTY), you need country(s) whose prefix(s) is *y*, eg, SENEED/RTTY YA.
SE/NEED/*x*/BAND =(*y*)*z*	Store in database that in mode *x* (where *x* equals CW, SSB or RTTY) on frequency band(s) *y*, you need country(s) whose prefix(s) is *z*, eg, SE/NEED/RTTY/BAND =(10) ZS9.
SE/NOA	Indicate that your computer/terminal is not ANSI-compatible.
SE/NOH	Indicate that you are not in your shack.
SE/Q *x*	Set your QTH as location *x*.
SHOW	Display requested information.
SH/A	Display names of files in archive file area.
SH/B	Display names of files in bulletin file area.
SH/C	Display physical configuration of cluster.
SH/C *x*	Display station connected to node whose call sign is *x*.
SH/CL	Display names of nodes in clusters, number of local users, number of total users and highest number of connected stations.
SH/COM	Display available show commands.
SH/DX	Display the last five DX announcements.
SH/DX *x*	Display the last five DX announcements for frequency band *x*.
SH/DX/*n*	Display the last *n* DX announcements.
SH/DX/*n x*	Display the last *n* DX announcements for frequency band *x*.
SH/FI	Display names of files in general files area.
SH/FO	Display mail-forwarding database.
SH/H *x*	Display heading and distance to country whose prefix is *x*.
SH/I	Display status of inactivity function and inactivity timer value.
SH/LOC	Display your station's longitude and latitude.
SH/LOC *x*	Display the longitude and latitude of station whose call sign is *x*.
SH/LOG	Display last five entries in cluster's log.
SH/LOG *n*	Display last n entries in cluster's log.
SH/M *x*	Display MUF for country whose prefix is *x*.
SH/NE *x*	Display needed countries for station whose call sign is *x*.
SH/NE *x*	Display stations needing country whose prefix is *x*.
SH/NE/*x*	Display needed countries for mode *x* where *x* equals CW, SSB, or RTTY.
SH/NO	Display system notice.
SH/P *x*	Display prefix(s) starting with letter(s) *x*.
SH/QSL *x*	Display QSL information for station whose call sign is *x*.
SH/S x	Display sunrise and sunset times for country whose prefix is x.
SH/U	Display call signs of stations connected to the cluster.
SH/V	Display version of the cluster software.
SH/W	Display last five WWV propagation announcements.
TALK	Talk to another station.
T *x*	Talk to station whose call sign is *x*. Send <CTRL-Z> to terminate talk mode.
T *x y*	Send one-line message *y* to station whose call sign is *x*.
TYPE	Display a file.
TY/*x y*	Display file named *y* stored in file area named *x*.
TY/*x*/*n y*	Display n lines of file named *y* stored in file area named *x*.
UPDATE	Update a custom database.
UPDATE/*x*	Update the database named *x*.
UPDATE/x/ APPEND	Add text to your entry in the database named *x*.
UPLOAD	Upload a file.
UP *x*	Upload a file named *x*.
UP/B *x*	Upload a bulletin named *x*.
UP/F *x*	Upload a file named *x*.
WWV	Announce and log WWV propagation information.
W SF=xxx, A=yy, K=zz,a	Announce and log WWV propagation information where xxx is the solar flux, yy is the A-index, zz is the K-index and a is the forecast.

21.250 MHz. I can add this information to the network by sending: **SV3AQR DX 21.250.**

Within seconds the spot will appear on the screens of everyone else who is connected to the network.

One question that comes up often is, "How do I know if a DX station is worthy of an announcement on the *Cluster*?" After all, what's choice DX to one ham is garden variety to another. My advice is to observe the activity on your network before you start posting spots of your own. On some networks, anything goes. Other networks, however, concentrate on rarer contacts. You might receive a sarcastic response if you post a spot for, say, a French station on 20 meters from the east coast. (A contact with France is not rare DX to most hams.)

In either case, make your spots accurate. Be sure you've entered the call sign and frequency correctly. If the DX station is working split, say so. Just a brief comment such as "LISTENING UP 30" is sufficient.

If there is a hot contest in progress, you'll know right away when you connect to the *PacketCluster* (see Figure 4). Just the nature of the spot listings will tip you off. In addition, the node may send a sign-on message telling you that the network is in the "contest mode." This means that certain *PacketCluster* functions are disabled during the contest. Generally speaking, you should only post information about contest contacts when the network is in this mode.

Above and Beyond

PacketClusters have other useful features. You can use the **DIR** command to see a list of bulletins just as you would on a packet BBS. Using the **R** (READ) command will allow you to read any bulletin you wish. Unlike packet bulletin boards, however, *PacketClusters* can relay bulletins and messages only within the networks they serve.

As I'm watching the *PacketCluster*, I see a spot for a DXpedition on Sable Island. Hmmm... I could use that one! I wonder what bands would provide the best propagation from my location to Sable Island? Why not ask the *PacketCluster*? All I have to do is send: **SHOW/M CY0.**

"M" stands for *maximum usable frequency* (or *MUF*) and CY0 is the call sign prefix for Sable Island. Here's how the *PacketCluster* responds:

Sable-Is Propagation: Flux: 137 Sunspots: 90 Rad Angle: 29

Table 2

Additional SHOW Commands

Note: Some of these commands may not work on your local *PacketCluster* network.

Command	*Definition*
SHOW/BUCKMASTER	Buckmaster US call-sign list
SHOW/RUMORS	Anything and everything!
SHOW/OBLAST	Russian oblast information
SHOW/PREFIX	Listing of countries and zones by prefix
SHOW/ALLOC	ITU allocation table
SHOW/STARS	Planetarium data
SHOW/BAND	Frequencies available for each class of license
SHOW/ZONE	Listing of countries and zones by prefix or zone
SHOW/DXNODES	Database of known DX *PacketCluster* nodes.
SHOW/BUREAU	World and Russian Oblast QSL bureau addresses
SHOW/CONTEST	Calendar and info on DX-related contests
SHOW/COORD	Longitude and latitude of US/DX locations
SHOW/DEALER	Amateur Radio dealer telephone/FAX directory
SHOW/DXCC	ARRL DXCC Country List
SHOW/FCC	FCC Field Office Directory and FCC Info Guide
SHOW/FLUX	Historical sunspot/solar flux data and glossary
SHOW/IRC	Required IRCs for QSL returns and IRC/postal info
SHOW/NCDXF	Information on the Northern California DX Foundation
SHOW/PUB	Directory of Amateur Radio publications from various countries
SHOW/QSLREC	Information on QSL cards received by users
SHOW/RULES	FCC Rules and Regulations Part 97
SHOW/IOTA	Islands on the Air (Award)
SHOW/TODAY	Did you know? Events of yesteryear!
SHOW/COUNTY	Directory of US counties
SHOW/EXTRA	Element 4B Extra Class Q&A
SHOW/INFO	International Q Signals, ARL Messages
SHOW/MIC	MIC wiring data prepared by KC4LWI
SHOW/LADDER	*PacketCluster* DXCC "Ladder of Success"
SHOW/SWL	Shortwave Listeners Guide

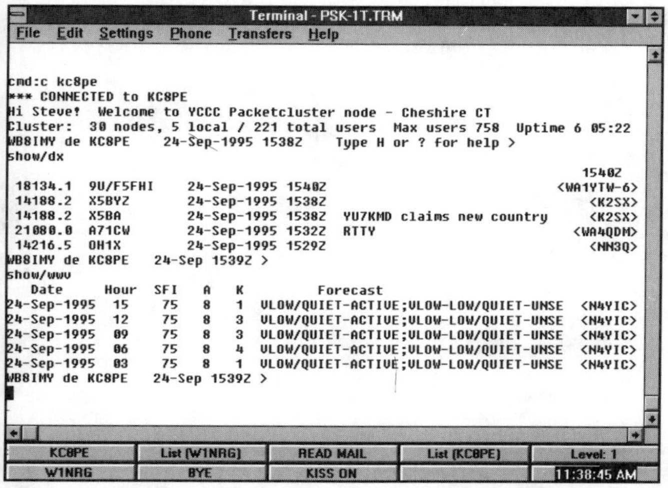

```
Terminal - PSK-1T.TRM
File  Edit  Settings  Phone  Transfers  Help

cmd:c kc8pe
*** CONNECTED to KC8PE
Hi Steve!  Welcome to YCCC Packetcluster node - Cheshire CT
Cluster:  30 nodes, 5 local / 221 total users  Max users 758  Uptime 6 05:22
WB8IMY de KC8PE    24-Sep-1995 1538Z    Type H or ? for help >
show/dx
                                                                 1540Z
 18134.1  9U/F5FHI    24-Sep-1995 1540Z                       <WA1YTW-6>
 14188.2  X5BYZ       24-Sep-1995 1538Z                           <K2SX>
 14188.2  X5BA        24-Sep-1995 1538Z  YU7KMD claims new country  <K2SX>
 21080.0  A71CW       24-Sep-1995 1532Z  RTTY                    <WA4QDM>
 14216.5  OH1X        24-Sep-1995 1529Z                           <NN3Q>
WB8IMY de KC8PE   24-Sep 1539Z >
show/wwv
   Date      Hour  SFI   A   K          Forecast
24-Sep-1995   15    75   8   1  VLOW/QUIET-ACTIVE;VLOW-LOW/QUIET-UNSE  <N4YIC>
24-Sep-1995   12    75   8   3  VLOW/QUIET-ACTIVE;VLOW-LOW/QUIET-UNSE  <N4YIC>
24-Sep-1995   09    75   8   3  VLOW/QUIET-ACTIVE;VLOW-LOW/QUIET-UNSE  <N4YIC>
24-Sep-1995   06    75   8   4  VLOW/QUIET-ACTIVE;VLOW-LOW/QUIET-UNSE  <N4YIC>
24-Sep-1995   03    75   8   1  VLOW/QUIET-ACTIVE;VLOW-LOW/QUIET-UNSE  <N4YIC>
WB8IMY de KC8PE   24-Sep 1539Z >

KC8PE | List (W1NRG) | READ MAIL | List (KC8PE) | Level: 1
W1NRG | BYE | KISS ON | | 11:38:45 AM
```

Figure 3—When I connect to my local *PacketCluster* node (KC8PE), I usually ask for a list of the most recent DX spots. I follow up with a request for WWV solar activity reports.

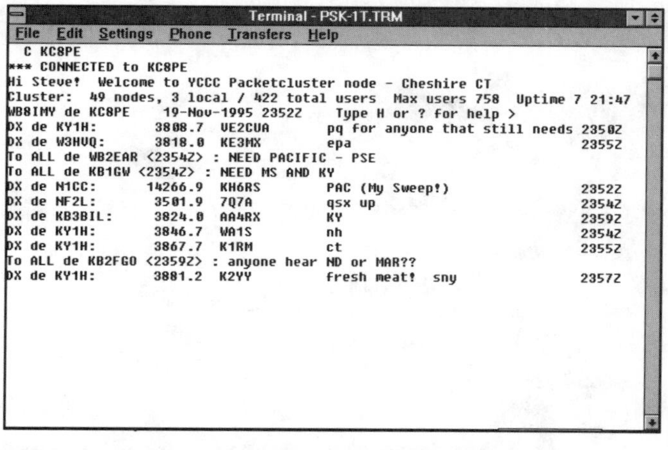

```
Terminal - PSK-1T.TRM
File  Edit  Settings  Phone  Transfers  Help
 C KC8PE
*** CONNECTED to KC8PE
Hi Steve!  Welcome to YCCC Packetcluster node - Cheshire CT
Cluster:  49 nodes, 3 local / 422 total users  Max users 758  Uptime 7 21:47
WB8IMY de KC8PE    19-Nov-1995 2352Z    Type H or ? for help >
DX de KY1H:       3808.7  VE2CUA       pq for anyone that still needs 2350Z
DX de W3HVQ:      3818.0  KE3MX        epa                            2355Z
To ALL de WB2EAR <2354Z> : NEED PACIFIC - PSE
To ALL de KB1GW <2354Z> : NEED MS AND KY
DX de N1CC:      14266.9  KH6RS        PAC (My Sweep!)                2352Z
DX de NF2L:       3501.9  7Q7A         qsx up                         2354Z
DX de KB3BIL:     3824.0  AA4RX        KY                             2359Z
DX de KY1H:       3846.7  WA1S         nh                             2354Z
DX de KY1H:       3867.7  K1RM         ct                             2355Z
To ALL de KB2FGO <2359Z> : anyone hear ND or MAR??
DX de KY1H:       3881.2  K2YY         fresh meat!  sny               2357Z
```

Figure 4—Here's a snapshot of my local *PacketCluster* network during the 1995 ARRL November Phone Sweepstakes. Notice that several spots have been posted for "rare" sections.

Dist: 995 km   Hops: 1

MUF (90%): 9.0 (50%): 10.7 (10%): 13.1

The MUF calculations tell me that propagation to the island is most reliable at 9 MHz (90%), but drops rapidly as the frequency increases. The 30-meter band (10 MHz) would be the best bet.

DX PacketClusters also allow you to chat with other stations in the network. You can send a simple greeting by using the **TALK** command:

TALK KX4V Hello, Rick. Nice job working 5U7M!

If Rick wants to talk to me at length, he can use the same TALK command to establish a link between our stations. The *PacketCluster* will continue to show us new DX sightings as they appear, but everything we type will be sent to each other.

If you want to see how the *Cluster* network is configured and who's connected to it, just send the **SHOW/C** (show configuration) command. The *PacketCluster* will respond with a complete list of every station connected to the network grouped by the node they're using. Here's a typical example:

PacketCluster Configuration:

Node	Connected stations
KC8PE	N1API KS1L K1WJL KC1SJ WB8IMY
W1RM	(WB1AIU) NTØZ KA1BSA AB1U KB1LE N1JBH N1GLA KG1D-1 (NJ2L) (NA1I) KB1HY K1FRD NX1L KB1BE W1CKA KB1CQ (W1GG) K1ZJH WB1GUY K1KI
K2TR	KA2EXB (K2QE) N2JJ NJ1F K2VV (KB2HUN) K2ONP KQ2K WK2H KA2HTU WS2U N2EKU

You can send TALK messages, or enter into a conversation, with any ham on the list. The only exceptions are call signs in parenthesis. These hams are connected to the *Cluster*, but they're away from their keyboards temporarily. How does the system know this? Actually, it doesn't—unless you tell it. (Yes, there is a command for this, too!) And like packet bulletin boards, you don't need to send the full command every time. Instead, you can use the abbreviated form such as **SH/DX** rather than **SHOW/DX**.

There are many other functions available, depending on the sophistication of your local network. See Table 2 for some of the more versatile and interesting SHOW commands.

Where's My *PacketCluster*?

There are *PacketCluster* networks in most urban areas, and even in some lightly populated regions of our country. Ron Rueter NV6Z, maintains a list of active *Cluster* networks, but it's too large to publish here. If you have Internet access, you can obtain the list from the ARRL InfoServer. Simply address an e-mail message to: **info@arrl.org**. Leave the subject line blank. In the body of your message enter the following on separate lines:

SEND CLUSTER.TXT

QUIT

The server will send you the complete list right away. You'll also find CLUSTER.TXT on the ARRL *Hiram* BBS at 860-594-0306. List updates can be found at the following Internet ftp site **pinsight.com** in the directory /pub/K6PBT.

And what if there are no *PacketClusters* in your area? Consider starting your own! You can obtain more information about *PacketCluster* software from the manufacturer: Pavilion Software, 8 Mount Royal Ave, Marlborough, MA 01752. Enclose a self-addressed, stamped envelope. QST

Stan Horzepa, WA1LOU*

Chasing DX with *APRS*

Two meters is open this morning. There is a tropospheric (or "tropo") inversion along the East Coast that is enhancing propagation out to a range of 100 to 200 miles. You would never know this by monitoring the 2-meter SSB calling frequency (144.200 MHz) as there is nary a signal to be heard there. So, how do I know that 2 meters is open? *APRS* is showing the way.

Long before the marriage of home computers and Amateur Radio, my main interest was in the VHF/UHF world. One of my first radios was a Heathkit 2-meter "Benton Harbor Lunchbox," also known as a "Twoer." It was a one-channel AM transceiver with a superregenerative receiver. "Super" was a misnomer. It certainly was not of Clark Kent's lineage as it left a lot to be desired in the selectivity department. It was so nonselective that one strong signal seemed to fill the whole 2-meter band!

But, because of my limited college income (or should I say "superincome"), I used to haul my Twoer around in my car looking for the high spots in New Haven County to work 2-meter DX. It was tough going, but I managed to work a handful of states because the receiver was sensitive, albeit nonselective.

Then came FM, but I won't get into that (if you're interested, see "FM/RPT" in *QST* between June 1979 and March 1990). Anyway, I was bit by the VHF/UHF bug a long time ago and I still chase 2-meter DX whenever it is available.

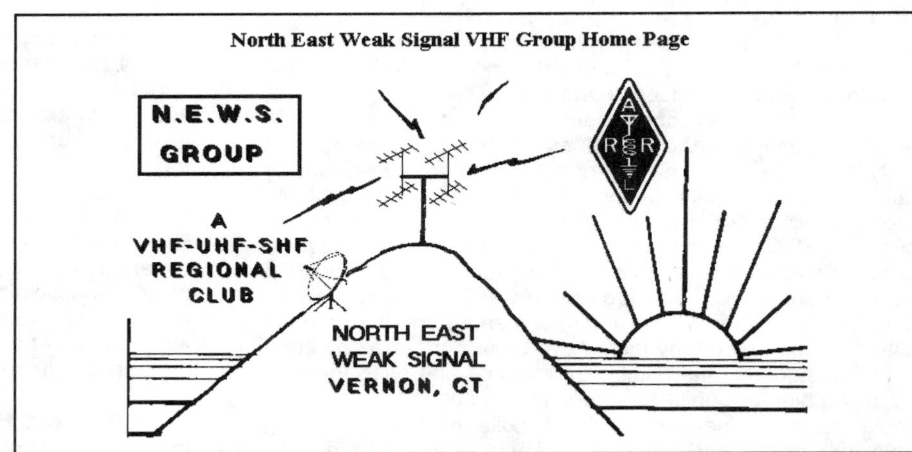

The North East Weak Signal VHF Group home page.

Propagation Tool Time

When I started playing with WB4APR's Automatic Packet Reporting System (*APRS*), I thought that it might be a suitable tool for chasing 2-meter DX. *APRS* is software that graphically represents the position of mobile and stationary packet-radio stations and other objects on a map displayed on your computer monitor. Most *APRS* activity occurs on the same channel (for example, 145.79 MHz on 2 meters) and, as a result, if you continuously monitor that channel, you can tell that the band is open when stations begin appearing on your map that don't normally appear there.

I was sold on using *APRS* as a 2-meter propagation alerting mechanism when one afternoon last July, stations in Georgia started popping up on my *APRS* map. I guessed that sporadic-E propagation was in the works, so I switched to SSB and worked a number of stations in Florida for my 26th state on 2 meters.

The outer limits of my *APRS* coverage area during normal 2-meter propagation conditions are WA2JNF in Brooklyn, New York, 85 miles to my southwest and W1TDG in Hinsdale, New Hampshire, 90 miles to my northeast. However, this morning, my computer monitor displayed stations such as WA1YKN on Cape Cod, 135 miles to the east, NR1N in Warner, New Hampshire, 133 miles to my northeast, and N2MSM on the South Jersey shore, 174 miles to my southeast, indicating that there was tropo inversion propagation afoot.

Wheat vs Chaff

To use this tool correctly, you must interpret what your *APRS* map is displaying. On a typical day, an *APRS* map displays a lot of stations within a 150-mile radius of your station as well as pockets of activity scattered across the continent. Such a display does not necessarily mean that the band is open.

Most of the stations within a 150-mile radius (give or take mileage depending on the lay of your *APRS* LAN) are displayed on your map because *APRS* digipeaters are relaying their positions to your station. Similarly, the stations beyond the 150-mile radius are likely to be HF *APRS* stations whose positions are being relayed to 2 meters by an *APRS* HF gateway station on your *APRS* LAN. In any case, you are not receiving most of these stations directly; repeaters and gateways are doing the work.

Only when you receive a distant station directly is it an indication of unusual propagation conditions. So, how do you differentiate the directly received stations from the digipeated and gatewayed stations?

If you use the *DOS* version of *APRS*, you can filter all but the direct stations by invoking the Controls-Filters-Direct series of commands. Once invoked, *APRS* only displays the stations you hear directly. There is no equivalent for the Controls-Filters-Direct series of commands in *MacAPRS* or *WinAPRS*, so differentiating between the direct and digipeated/gatewayed stations on a Macintosh or *Windows* platform requires more work.

After you have used *MacAPRS* or *WinAPRS* for a while, you get a feel for which stations are within your normal limits, as well as the HF stations that are being gatewayed to your LAN. When an unfamiliar station appears on your map, you can check how you are receiving it by using your mouse to double-click its icon to display its path. For example, today the path of K1HJC in Candia, New Hampshire, appears as K1HJC>APRS, NR1N-2,WIDE.

This indicates that I am receiving K1HJC directly. An asterisk in the path would indicate otherwise. For example, if an asterisk followed NR1N-2 or WIDE, it would indicate that I was receiving K1HJC via digipeater NR1N-2 or via a digipeater with an alias of WIDE. Similarly, if GATE* appeared in the path, it would indicate that the station was being gatewayed to me.

That explains how you can use *APRS* as a VHF/UHF propagation tool. Now let me go and look at my *APRS* map. Maybe I can work state number 27!

Wireless WWW Page of the Month

While I am on the subject of VHF/UHF DXing, it is appropriate that the Wireless World Wide Web page for this month is the North East Weak Signal VHF Group home page at **http://uhavax.hartford.edu/newsvhf**. This page has a fine selection of files and software related to VHF/UHF DXing and contesting. It also has a long list of links to other VHF/UHF-related sites throughout the world.

*One Glen Ave
Wolcott, CT 06716-1442
e-mail **stanzepa@nai.net**

Pactor-II

Impressions and Update Information one year after

By Dr. Tom Rink, DL2FAK & Dipl.-Ing. Martin Clas, DL1ZAM
Roentgenstrasse 36, D-63454 Hanau, Germany

I. Introduction

PACTOR-II was introduced together with the new multimode DSP controller PTC-II about one year ago. It includes several advantages that cannot be found in any other digital mode, such as a powerful convolutional code with a real Viterbi decoder to increase the robustness with weak signals, or a newly developed on-line data compression system (Pseudo-Markow Coding), which, along with the run-length encoding, roughly doubles the effective throughput. Therefore, the interest in the PTC-II units has always been tremendous, which caused the manufacturer to run out of stock several times. Up to now, more than 1000 modems are sold in all continents, about half of them in the commercial market, and the demand is still growing fast. Many tests have been performed by commercial customers, like relief organizations and even the military, using highly sophisticated equipment such as ionospheric simulators, to compare all available digital modes concerning bandwidth, speed and robustness. They, as well as the radio amateurs who tested it with real band conditions, clearly proved PACTOR-II to be the fastest and also the most robust digital narrow band mode for data transmission on short wave available at the moment. Information can still be transferred when a signal is 18 dB below the noise level. This, for example, allowed a QSO between Germany and a mobile Australian station with absolutely inaudible signals and only 16 mW of HF power. In good conditions, PACTOR-II easily exceeds an effective throughput of 1000 bits per second. The required signal bandwidth is just 450 Hz (at minus 50 dB), regardless of the actual modulation form and the transferred speed. Unlike other DPSK modes, PACTOR-II tolerates a high frequency offset when connecting and also a high frequency drift in an established link, as newly developed frequency and phase tracking algorithms are used, which still work in the above described borderline conditions. Initial frequency offsets of up to +/-80 Hz are automatically compensated by the PTC-II without the need of any user access. This process can be observed on the multicolor tuning display. For operating PACTOR-II, high tuning accuracy or frequency stability of the transceivers are hence not required. Detailed information on the technical basics, the PACTOR-II protocol and the PTC-II hardware can be found in a four part's series on PACTOR-II, which was published in the January to April 1995 issues of the Digital Journal.

The PTC-II is much more than just a multimode modem for data communication on short waves. It features the powerful processor MC68360 which also includes 4 SCC's implemented as a RISC system. Four different communication channels are thus simultaneously supported by the unit. The 60 MHz version of the special DSP 56156 is used for the HF port and also for many other tasks, such as audio processing, etc. The firmware is always being expanded in order to improve the system and to add new modes and functions. All new firmware releases are available for free, and can be downloaded from the SCS mailbox, which is available 24 hours a day at (+49)-6184-900427. Since the PTC-II came into the market about one year ago, many new features have been implemented, like a new and for DSP operation optimized CW-decoder, the above mentioned audio DSP filter, an automatic output power control in order to match the needs of the current link, and a complete remote control capability for Icom, Kenwood, SGC and Yaesu radios using the special transceiver control port. This last mentioned feature, for example, allows one to scan pre-defined frequencies without the need of an external computer. All transceiver control commands as well as the required handshake procedures are automatically generated by the PTC-II. As the voltage levels are compatible in most cases, an additional interface between the PTC-II and the transceiver is usually not required.

A lot of new features will still be added this year, for example Packet Radio, a complete host mode and the support of additional modes, like FAX and SSTV. Explaining the details of all possible applications would surely exceed the range of this article, therefore we have to restrict to three of the novelties, which were already added last year: The audio processing function, the transceiver control option and the enhanced CW operation.

II. DSP Audio Filter Operation

Independent of its function as a DSP multimode controller, the PTC-II can be used for special processing and filtering of any audio signal. It is thus also a comfortable stand-alone DSP filter, suitable for SSB operation, CW listening, and many other tasks. The high processing power of the PTC-II has proven to be very advantageous for this audio filter function. In comparison with the usual simpler and cheaper DSP audio denoiser units, much more computing effort can be used to obtain an optimum result. The commands controlling the DSP audio filter function can be found in a special sub-menu called 'Audio'.

The AF signal is supplied to the PTC as usual via the HF radio connector, and thus no changes of wiring compared to the 'normal' RTTY/AMTOR/PACTOR operation is required. The processed and filtered AF is available at another pin of that connector and additionally at the Mini-DIN-connector (8-pin TRX-Remote-Control). All functions of the Audio sub-menu that evaluate the AF input signal, use a 4-stage signal level matching (22 dB control range) for the 16-Bit A/D converter, in order to keep the quantisation effect as low as possible and to allow a large effective dynamic range. The PTC-II therefore adjusts itself automatically to the average signal level delivered by the transceiver. The maximum level of the AF output signal is +/-500 mV. For the first test, a 600 Ohm earphone can be connected directly to either of the two AF outputs. Nevertheless, a small AF amplifier with volume control is recommended to be used, which enables comfortable speaker operation as well as the connection of any kind of earphones. Such a unit, which additionally provides remote controllable, bi-directional digital and analog user ports, will be available as an option later this year (see below). As the unregulated supply power of the PTC-II is also available at the Mini-DIN-connector, the unit can directly be connected there, and does not need any additional wiring.

Here is a short explanation of the most important commands of the Audio sub-menu. 'Notch' activates the automatic N-times notch filter. All systematic signals are heavily attenuated. The used algorithm leads to considerably less signal distortion of speech signals compared to simpler DSP notch filters. 'Peak activates the automatic N-times peak filter. This may be considered as a phase-linear auto-correlation filter, of a very high order and large dynamic range. This filter enables slow CW-signals within the SSB bandwidth to be found that are BELOW the level discernible by the human ear. The filter algorithm puts a very narrow band filter on all systematic signal components. Uncorrelated noise is heavily attenuated. The active mode is always displayed on the dot-matrix display of the unit as well as on the connected terminal. The command 'CWfilter', activates the CW filter, using a center frequency and bandwidth that can be freely defined. The filter is designed as FIR with a linear phase-change, so that even with a bandwidth of 30 Hz it does not ring. The transfer function is not designed for maximum slope steepness, but a shape leading to a signal easily readable for the human ear, and the best obtainable signal to noise ratio.

III. Transceiver Remote Control

The commands required for the transceiver control function can be found in the sub-menu 'TRX', which is entered using the 'TRX' command (without argument). This command may also contain an argument, where all applicable commands from the TRX-menu are allowed. In this case, the PTC will carry out the command without switching into the TRX sub-menu. The control commands can thus considered to be 'fed through'.

Example: 'TRX Frequency 14079.0 <Enter>'

changes the frequency of a connected transceiver directly to 14079.0 kHz - without having to enter the TRX-menu. In the following, we introduce the most important commands of the TRX-menu:

The 'Channel' command allows up to 16 channels to be defined. Every channel consists of a channel number, its frequency (in kHz), the scan status and an optional short comment. Information on the current entries can be obtained using the 'List' command. Such a frequency list could look like the following example:

CHANNEL-LIST: ========	Ch	Frequency (kHz)	Scan	Comment
	1:	14079.000	YES	DL2FAK Main QRG on 20 m
	2:	14077.000	NO	Test QRG
	3:	3584.000	YES	DL2FAK on 80 m

The 'Channel' command (without argument) behaves similar to the List command. All user defined channels are listed. If the Channel command is followed by ONE argument, consisting of a number between 1 and 16, the PTC switches the connected transceiver to the frequency of the given channel. If, for instance, the command 'CH 3 <Enter>' is given, then (considering the above example) the transceiver would be switched to 3584.000 kHz. The definition of a channel is carried out by putting two or three arguments after the 'Channel' command.

Example: 'C 10 14076.5 EA5FIN STBY FREQUENCY <Enter>'

defines the frequency 14076.5 kHz as channel number 10, with the comment 'EA5FIN STBY FREQUENCY'. The comment does not contain essential information and may be omitted. The frequency input is always in kHz. The decimal point after the megahertz position is optionally allowed (e.g. '14.076.50'). The last decimal point is processed as a kilohertz decimal point. There are up to three positions allowed after this decimal point. A frequency accuracy of 1 Hz is thus anticipated, which, however, is not supported by some transceivers. The frequency input 0 kHz erases the channel from the frequency list.

The 'TYpe' command is used to set the transceiver type for the configuration of the PTC-II interface. There are up to three arguments allowed. The first one defines the transceiver (currently Icom, Kenwood, SGC and Yaesu, but this list will be extended). The second value is the baud rate. When using Icom equipment, the PTC requires a number as the third argument (maximum 2 figures) that represents the transceiver address number. With Kenwood and Yaesu equipment, the third argument is the VFO number (A or B) that should be addressed by the PTC. The 'Scan' command has two different functions: An argument 1 or 0 switches the scanner on and off, respectively. It is thus the 'main switch' of the scanner. If, as argument, the word 'Channel' (minimum abbreviation: 'C') follows, a channel number of the frequency list can be entered to toggle the scan status between 'YES' and 'NO' for the respective channel. This feature hence allows a channel to be scanned or skipped. The 'DWell' command sets the dwell time of the PTC scanner on each channel in 100-ms steps. A dwell time of 30, for example, means that the scanner will pause on each channel for exactly 3 seconds. The 'Wait' command defines the time (in seconds) that the scanner waits at the respective channel after a disconnect occurred before the scanner starts again. The 'Offset' command applies a frequency offset to EVERY channel of the channel list before it is sent to the transceiver. The valid range is -5.000 to +5.000 kHz. This allows the PACTOR mark frequency to be entered, even whilst in SSB mode. If, for instance, low-tones are used (1200/1400 Hz) and USB position, the transceiver is set on 14077.60 kHz in order to transmit the mark frequency of 14079.00 kHz. As the transceiver displays the frequency of the (imaginary) carrier, the frequency of the audio-mark-tone (1400 Hz) must be added to the carrier frequency in USB, to get the actual mark frequency. If, on the other hand, any mark frequency is taken from a BBS list, the mark tone frequency must be subtracted in order to find the correct frequency to tune the SSB transceiver to. If the offset value is defined as -1.4 kHz, the PTC-II carries out the required frequency correction for the mark frequency automatically. It is thus only necessary to enter the desired mark frequency, and the correct offset is automatically applied. With regard to the first above mentioned example, one can give the command 'F 14079.0 <Enter>'. The PTC-II then switches the transceiver to 14077.6 kHz, which automatically leads to the required transmit and receive mark frequency of 14079.0 kHz.

The 'Down' and 'Up' commands allow the microphone down or up key to be activated (simulated by an FET switch), which is connected to the corresponding pin of the HF transceiver socket. This way it is possible to adjust the frequency without accessing the serial interface. As argument following the command, a number between 1 and 60000 is entered. This represents the number of key pulses initiated by the PTC-II. With no argument, a single pulse is given. 'Ptime' sets the time (in milliseconds) for these up and down keying pulses. For example, a Ptime-value of 50 means that the respective switch in the PTC-II is closed and opened for 50 ms per pulse.
The commands 'Dump' and 'Transfer' are used to send an ASCII string or HEX dump to the transceiver. This way, any command or remote control string supported by the transceiver can be passed through the PTC-II, for example from an external control software.

A description of the full TRX command set as well as more details on the above mentioned commands can be found in the update information file, which is available at the SCS phone mailbox or in the Packet Radio and PACTOR network.

IV. Improved CW Operation

The CW routine is implemented using a highly sophisticated DSP technique. The demodulator utilizes the much talked about technique of the auto-correlation filter in the Audio menu (see above). This enables even weak signals to be reliably detected without any tuning problems. (An ideal filter for a CW signal with a speed of 60 cpm exhibits a 0/0 bandwidth of only 20 Hz that - with conventional methods - requires extremely exact and stable tuning for good results.) The auto-correlation method is also the basis of the AGC used in the CW demodulator, with a dynamic range of approximately 40 dB. The AGC allows a constantly good receive performance, independent of the audio input signal level.

For easy and fast accessing operation, the CW terminal offers the use of a number of so-called hotkeys: Pressing the Break-In key twice switches between direct transmission of the keyboard input (immediate transmission mode), and a delayed switch over (delayed transmission mode). This delayed switching allows text to be 'typed ahead' whilst reading the other QSO partner's transmission. The text in the buffer can then be transmitted by pressing the Break-In-Character once. The transmission is then only blocked again when no characters are transmitted for 6 seconds. This way, the user may continue to write after the buffered text is sent, without having to press any other key. The renewed blocking of the transmitted text (it is then being redirected into the buffer instead of being immediately transmitted) is shown by the PTC with the message ">>>" written into the Echo-Window. After switching to the CW terminal, one is always in the direct transmission mode, and the automatic speed adjustment mode is activated. '<CTRL-F>' switches between automatic and manual RX speed adjustment (fixed speed mode). On switching between these two possibilities, the present speed is taken without change. Thus the automatic adjustment can be left on for a while to detect the correct speed, then '<CTRL-F>' can be pressed to keep this setting. '<CTRL-U>' ('Up') increases and '<CTRL-D>' ('Down') decreases the decoder speed by 1/16 of the actual value. This is important when the automatic RX speed adjustment is turned off. The operation with a fixed speed has definite advantages in weak signals and in signals with heavy fading. The decoder withstands speed errors of 40 percent without any problem. Hence even with a fixed speed setting, virtually no reading errors are found. The CW speed, regardless of whether it was automatically detected or manually set, is shown at the dot-matrix-display.

V. A Look to the Future

Packet Radio and the host mode are certainly the next features to be added to the PTC-II. For the Ham Radio exhibition in Friedrichshafen/Germany, which takes place in June this year, the 1200/2400 Baud as well as the 9600 Baud plug-in modems are planned to be available, together with the corresponding firmware update. At the same time, the above mentioned Remote-Control-Amplifier-Unit (RCU) for the PTC-II will be available. This small box is connected to the PTC-II using the 8-pin Mini-DIN connector. It has two major features: At first, it provides an 8 W audio-amplifier with volume control, speaker and headphone connector to support and simplify the audio denoiser operation with the PTC-II. As a second feature, it provides a mixed-mode user-port interface to the PTC-II, accessible from the local terminal as well as from each communication channel (HF or Packet). It consists of eight digital inputs, eight digital outputs, eight analog inputs and eight analog outputs. Connected to any imaginable application, remote-control or remote data acquisition is possible using a subset of user-friendly commands. A detailed description of the RCU will be published separately.

(Note: please see the editorial comment regarding Pactor-II in the Last Word column—de N2HOS)

The HAL P38 DSP

and other thoughts

by Hal Blegen, WA7EGA
2021 Smythe Road • Spangle, WA 99031

The HAL P38: Digital Modes on a card

I'm a self-made advertising pariah. I have not been approached to do a product review since I submitted scientific evidence, a few years back, proving that on RTTY, a PK232 was only slightly more efficient than my dog Brownie who learned baudot by sleeping in the furnace room next to a Teletype model 15.

I don't like to work on computers. It has been months since I last pulled the cover on my 486. The anxiety I associate with trying to get WINDOWS to talk with all the ports and interrupts so that a mouse, modem, sound card, FAX, cd rom, printer, joystick and a pair PK232s will all work has me stressed to the point where the mere sight of a phillips screwdriver gives me diarrhea. It's no wonder that the UPS delivery guy thought I was acting a little strange. All that was visible to him was a mailing label indicating that the package came from HAL COMMUNICATIONS but to my, more experienced eye, what it said, DANGER, THIS PACKAGE CONTAINS A POISONOUS REPTILE!" My P38 HF RADIO DSP MODEM had arrived.

Installation instructions: Turn off the PC and stuff the board into any empty slot. Naturally, the pint-sized fan that my good friend Jeff Flashnerd mounted on my processor chip to solve the meltdown-lockup problem stuck up into the only open slot and the P38, not one of your whimpy little serial port boards, needed the whole slot. The first sunshine in my otherwise rainy life was, after shuffling the deck to get an open slot, the computer still worked! Unless your rig cannot use open collector switching for FSK keying, no switches or jumpers have to be configured to get the board to work.

While this is not about software but trying to talk about a software driven device without mentioning software is like describing a fish without mentioning water. One unique features of the P38 is the lack of E-PROMs. The basic run-time software for the microprocessor is downloaded each time you initialize the control program —it's just a file. This should significantly reduce the handling and distribution overhead of software upgrades, allowing the newest version can be downloaded from a BBS or an FTP site on the Internet.

HAL's software runs under plain-vanilla, DOS but for mouse-bigots there are several authors offering WINDOWS compatible, P38 software. The only compatibility problem I found was that different authors used unique names for the run-time files and some renaming was required (RAGCHEW worked fine until I installed EXPRESS at which point I had to reinstall RAGCHEW to get it to work again).

It was refreshing to run the install program and not have the screen all cluttered up with dire warnings about insufficient free RAM and EMM386 incompatibility. The programs don't require much disk space and, best news yet, THE P38 DOESN'T NEED EITHER AN INTERRUPT OR A COMM PORT! The hex memory address for I/O is user selectable so you can avoid conflict with common cards like Sound Blaster or the ever-popular Dental-Surgery adapter. I installed it and it worked, the first time!

The interface to the radio is dead simple. In recognition of the fact that most HF operators aren't smart enough to run AFSK without transmitting 5 khz worth of birdies, the P38 comes with a built-in FSK keying line. On most radios, unless you want to run CLOVER, all you need is FSK, Audio in, and PTT. I'll probably go down in flames for saying so, but the average human who ragchews and works a little DX needs CLOVER like a guy in rubber life raft needs a chainsaw. CLOVER is for adults.

Noticeably absent was any sort of scope driver output. The P38 operating screen displays a set of bar graphs for real-time tuning that takes about 10 seconds to confirm what a scope shows at a glance. Solid state tuning indicators annoy me so much that I have resurrected an old Flesher TU-170 just to drive a tuning scope.

The P38 does CLOVER, AMTOR, PACTOR, ASCII and BAUDOT. It doesn't do the G thing (You know, the one that was named by Dr. Ruth?). At Dayton somebody asked Bill Henry why the board didn't do PACKET. He just laughed and pointed out that, "The P38 was designed to work on HF."

PACTOR on the P38 was another happy experience. In fact, if they leave things along and don't invent 10 more modes to confuse the issue, FEC PACTOR may well replace BAUDOT. It's just a matter of getting a few of us old die-hards to buy some new equipment. With 5 watts out, I linked with a ZL (no timing problems on my elderly ICOM 751 using the default parameters) and once I changed from 250 to 400 hz bandpass, the P38 switched to 200 baud which boosted the throughput to about 10 pages of copy per minute and made an instant convert out of me. When I went to SSB filters at 2.8 khz, the DSP filters completely ignored another station that was in the passband even though the interfering signal was pumping my AGC to levels above the linked signal level!

A rumor about the P38 that peaked my interest enough to buy (I paid —no cozy deals), had to do with some supposed comparisons to HAL's, top-of-the-line ST-8000. The P38 was supposed to be "only 1 db down" from the 8000. I didn't know what that meant but that I thought I would find out.

Error correcting modes give perfect copy under varying conditions. The down side, since only one device can control a link, is that getting any useful data in real-time comparisons between boxes requires more than just a couple of radios. Rather than spend a lot of time with multimode boxes in ARQ LISTEN modes I figured that overall demod effectiveness was unchanged on all modes except CLOVER so I went back to the old standard, RTTY-BAUDOT to do my comparisons.

On VOLTA contest weekend, I connected a HAL ST8000, an AEA PK232, a DOVE-TRON MPC1000, and the P38 to a single, buffered audio source and, armed with VCR to record audio, went in search of the crummiest signals I could find. No surprise. The ST8000 still wins. The P38 was an improvement over my early-serial-number PK232's and held it's own against a 15-year-old Dovetron. On over-the-pole, warbly dx signals (20 meters) there wasn't enough difference between the 8000 and the P38 to cost me a single contest QSO. At the point where both units started taking hits, the ST8000 usually recovered several characters sooner than the P38 but I was impressed.

Next, I dropped down to 80 meters. Yeah, I know, NOBODY runs RTTY on 80 meters. On the low bands, propagation does some funny things. On an otherwise loud signal, the mark and space can fade separately. There is also a lot of pulse stretching that can be a challenge. I remembered that Mike, N7RY maintains a rig on autostart at 3612.5 and I started playing his message-of-the-day file. The MPC-1000 (an AM unit) and the ST-8000 (running FM) never missed a beat. The PK232 fumbled now and then but the P38 might as well have curled up next to the furnace and gone to sleep. Not good.

I would have expected HAL COMMUNICATIONS to know all about selective fading and multipath so Monday morning I was on the phone —amazing, straight through, first time, to Bill Henry. Since I've heard a lot of variations on the I'll-look-into-it routine. I didn't expect very much. When I saw the E-mail saying the engineering department had more-or-less had confirmed my findings and were working on the problem, I was sorta impressed. When about a month later a new version of the code showed up on the internet, I was really impressed.

There is a point to be made here that doesn't fit the format of a product review. As far as the ability to put copy on your screen, I don't see a lot of practical differences between multi-mode boxes, what you're buying is just different bells and whistles. You might as well just look at the price. What I do think is important, is the general attitude that I found when I called the manufacturer. "Yep," they said, "We agree with you and we can fix it." What a concept!

Nope, the P38 still cannot out perform the ST8000, especially on 80 meters, but then, what kind of dummy would actually expect a $400 board that comes with free software to beat a top-end, mil-spec demod that sells for $4000. Right now, the P38 DSP is an evolving product. The current version performs very well in its market niche and, as it looks to me, the hardware I am now using will be able to utilize future evolutions without having to dismantle my computer every time I upgrade. I like that.

See ya on the air, Hal WA7EGA

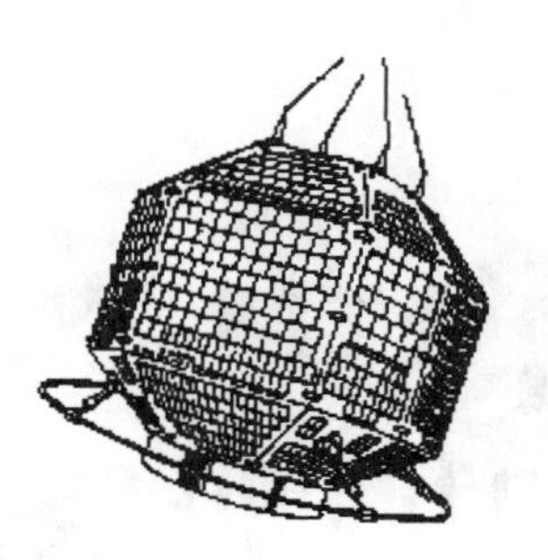

Fuji-OSCAR-29

Satellite Summary

Name: Fuji-OSCAR-29 aka Fuji-2 and JAS-2
Callsign: 8J1JCS
NASA Catalog Number: 24278
Launched: August 17, 1996
Launch vehicle: Japanese H-II No. 4
Launch location: Tanegashima Space Center of NASDA, Tanegashima Island, Japan
Weight: 50 kg
Orbit: Polar LEO (Low Earth Orbit)
Inclination:
Size: 44 cm wide x 47 cm high
Period:

Features:

- BBS Message System (digital store-and-forward)
- Analog Communications Transponder
- Attitude Control
- Digi-Talker
- Testing of newly developed solar cells in space

Beacon (100 milliWatt) Telemetry Format

435.795 MHz - CW
435.910 MHz - PSK digital - Digi-Talker

Digital Transponder - Mode JD (1 Watt)

- Uplinks: AFSK (FM) 1200 bps, AX.25, Manchester Encoded
  - 145.850 MHz
  - 145.870 MHz (the only 9600 bps uplink frequency)
  - 145.890 MHz
  - 145.910 MHz
- Downlink: BPSK 1200 bps or FSK 9600 bps
  - 435.910 MHz (also Digi-Talker frequency)

Analog Transponder - Mode JA (1 Watt)

- Uplink: 145.900 - 146.000 MHz
- Downlink: 435.800 - 435.900 MHz

References

- Steve Ford, WB8IMY, "JAS-2 In Orbit!," *QST*, Oct. 1996, p. 94.
- Fujio Yamashita, JS1UKR and Hideo Ono, JA1BU, "JAS-2 Comes to Life as FO-29," *The AMSAT Journal*, Vol. 19, No. 5, Sep/Aug 1996, p. 1.

WinAPRS™

Windows Automatic Position Reporting System
A Windows™ Version of APRS™

Mark Sproul, KB2ICI
sproul@ap.org

Keith Sproul, WU2Z
ksproul@noc.rutgers.edu

Abstract

WinAPRS is a Windows version of the popular *APRS*, Automatic Position Reporting System. WinAPRS is fully compatible with *APRS™*, The DOS version, and *MacAPRS™*, the Macintosh version. Due to the larger amounts of memory available in the Windows operating system, WinAPRS, just like MacAPRS has many additional features not available in the DOS version.

WinAPRS

WinAPRS is growing rapidly. Just like APRS and MacAPRS, the users are finding more and more things to do with this technology. We (Bob Bruninga and the Sproul Brothers)are committed to keeping the on-air protocols the same and are working with many different groups to expand and add many different capabilities to the APRS group of programs. One of the recent developments along these lines is a large interest from several National Weather Service groups across the country.

WinAPRS uses the exact same map files as MacAPRS, and will also use the map files from DOS APRS. Most of the source code of WinAPRS is the exact same code as MacAPRS, so it has been around for a few years, and has been thoroughly tested. See the discussion below about the development system used for MacAPRS/WinAPRS.

WinAPRS is a full Windows-95 32-bit application that follows the Windows User Interface Guidelines. It runs under Windows-95 and Windows-NT, and will run under Windows 3.1 and 3.1.1 if you have the Win32 DLLs installed that allow Win95 applications to run under the older versions of Windows.

History of APRS

1992

APRS™ was first introduced by Bob Bruninga, WB4APR, in the fall of 1992 at the ARRL Computer Networking Conference in Teaneck, New Jersey. [1]. We, (Mark and Keith) were at this conference and saw Bob's program. Keith commented that he wanted to do some of this, but when we asked how much a GPS (Global Positioning Unit) cost, we got an answer of $3000!. We decided to wait.

1993

APRS started gaining popularity. There were several articles in different magazines and many new uses for this growing technology. The article that caught a lot of attention was about using APRS to track the football from the Naval Academy to the Army-Navy game in Philadelphia. [2]

1994

In the fall of 1993, just about a year later, Keith started working on MacAPRS. [3] He contacted Bob Bruninga in February of 1994 and went to see him, with a working version of APRS that ran on a Macintosh. This version had many enhancements over the basic APRS features, including Call sign look-up from CD-ROM, and multiple maps open at the same time.

When Bob introduced APRS, all of his maps were made by hand! Keith, having had experience in college doing Cartography programming, refused to do maps by hand and did all of the maps for MacAPRS using USGS (US Geological Survey) map data, available on CD-ROM. Soon after Keith's visit to Bob, he started using the USGS CDs too. This improved the map quality greatly.

1995

By this time, APRS, and MacAPRS were becoming very popular and the uses of this technology had expanded much beyond the original concepts. The APRS programs have been used for Fox Hunting, Balloon Tracking, Weather Networks, DX Cluster monitoring, and many other applications. [4] [5]

At the Dayton Hamvention in April of 1995 Mark and Keith presented more and more of the fancy capabilities of MacAPRS. During 1995, we were invited to give talks at other hamfests and clubs in the New York/New Jersey/Connecticut area. During this time, one of the more common questions was "... do you have a WINDOWS version?..."

One of the more 'popular' features was the fact that MacAPRS did not really have any limitation as to the number of points that could be in a map. The DOS version, which when it first came out, was limited to 1,500 points had been upgraded so that it could handle 3,000 points, But the typical MacAPRS maps STARTED at 10,000 points, with some maps as large as 300,000 points. Other features that people were interested in that were not in the DOS version were the interface to the many different types of call-sign databases on CD-ROM.

At the Dayton Hamfest, we started getting more and more pressure from the Ham Radio community to do a Windows version. This PRESSURE got really severe at the ARRL DCC in Arlington, Texas.

When Keith got back from the ARRL DCC in Texas, we had long talks about doing a Windows version. Mark made the comment:

"I have never had so much peer pressure in all of my life..."

At this time, several critical items came together. CodeWarrior, the development system that the Sproul brothers used for MacAPRS came out with support for developing Windows programs on the Macintosh. Mark Sproul, who is porting MacAPRS to Windows finally succumbed to the pressure from APRS users. When these things happened, we determined that it was realistic to port the already developed Macintosh code to Windows and decided to do a Windows version of APRS. On September 15th, Keith went to down to see Bob Bruninga to discuss doing a Windows version. On September 16th, the following announcement was put up on the Internet:

MacAPRS™ for Windows
(WinAPRS™)
Automatic Position Reporting System for Windows

September 16, 1995
NORTH BRUNSWICK, NJ: Mark Sproul (KB2ICI) and Keith Sproul (WU2Z) authors of MacAPRS™, the Macintosh version of Bob Bruninga's (WB4APR) popular packet radio mapping system announced today that they will be porting their Macintosh version to Windows. This will be the official version and has the backing of Mr. Bruninga. The current plans are for beta release by Christmas 1995 and for the final release to be at the Dayton Hamvention in May of 1996.

APRS is a multi-faceted system used primarily within Amateur Radio for tracking many different types of things. APRS is used for tracking Weather, for tracking moving cars, boats, weather balloons, and many other things. It can also be used as Graphics Information System for many different aspects of Amateur Radio.

The original version of APRS was developed by Bob Bruninga, WB4APR, to run under DOS and was introduced at the 1992 ARRL Computer Networking Conferences. MacAPRS was released at the Dayton Hamvention in 1994.

The Macintosh version is written entirely in C and will port easily to Windows. Keith and Bob have worked hard at keeping the two versions compatible and by using all of the C code already developed for the Macintosh version, it will ensure complete compatibility on the Windows version. In addition, the two versions will use the exact same map file format so all of the wonderful maps that the Mac users have will be immediately usable by the Windows version.

When asked about future plans, Mark said, "When we finish with the Windows version, we are planning on doing an X-Windows version as well."

October 14, 1995

One day less than one month after deciding to do WinAPRS, we had the maps drawing on a Windows computer and put screen-dumps of these maps up on the Web for all to see.

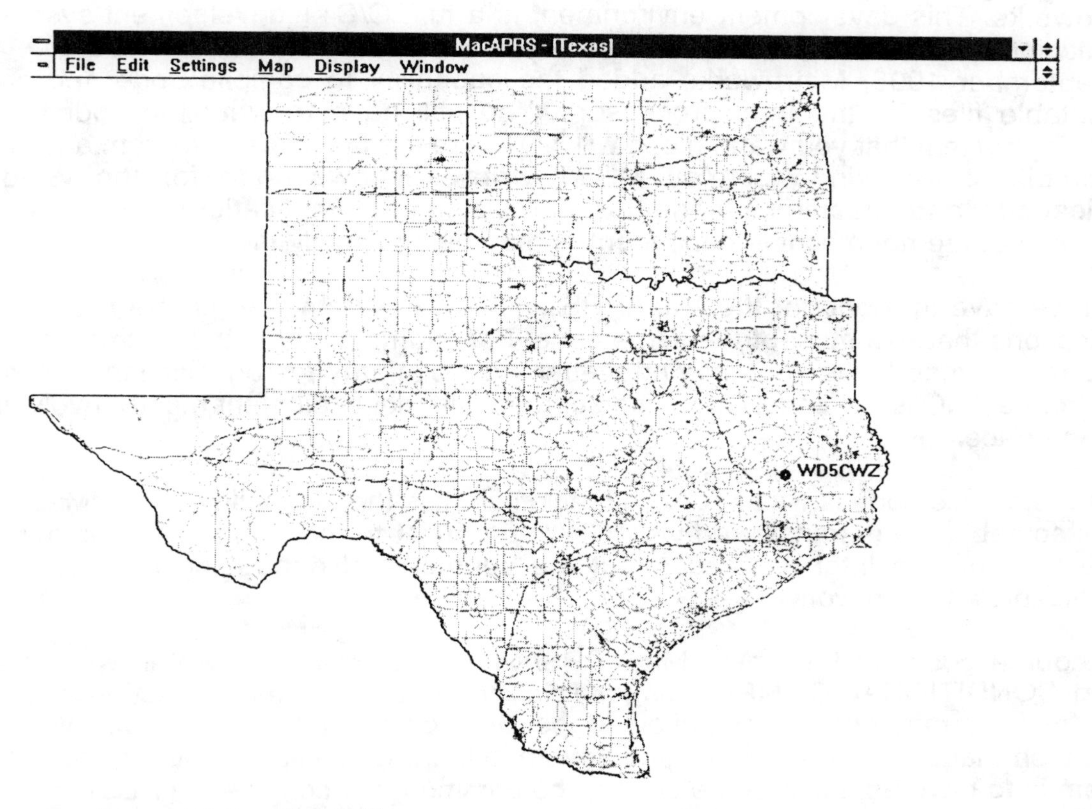

December 22, 1995

As promised in the original announcement, we released WinAPRS before Christmas. This release was to about 20 people.

January 28, 1996

We released a public beta version to the ham radio community. We showed WinAPRS publicly for the first time at the Wharton Hamfest near Chicago, Illinois.

May 1996

Again, as promised in the original announcement, we released WinAPRS version 1.0.0 at Dayton Hamfest 1996! This release had more features in it than we originally expected to have done at this time.

Development System of WinAPRS and MacAPRS

METROWORKS CODE WARRIOR

The Development system that we have been using for MacAPRS is Code Warrior by Metroworks. This development environment is a full C/C++ development system for the Macintosh. MacAPRS was written entirely in straight 'C', with no C++ at all.
In September 1995, Metroworks added the capability to compile code and create executable files for the Intel processors. You still have to write the code for the operating system that you want, i.e. it will NOT take the Macintosh program and simply re-compile it for Windows. You MUST write Windows code for the Windows applications and Macintosh code for the Macintosh applications. However, the routines that are not machine dependent end up being exactly the same.

What we have for done for the MacAPRS/WinAPRS system is to create two different applications that use most of the same code. For example, doing the math for drawing maps from a map file is the same no matter what platform it is on. Similarly, decoding data from a TNC is the same, etc. The source code that is different mostly involves the user interface.

All of the source code is written on the Macintosh. It is then compiled on the Mac. Then the executable file is transferred via TCP/IP-EtherNet to the Windows computer. The Code runs on the Intel processor, but the source-level debugging is done on the Macintosh via the network.

The source code for the entire MacAPRS/WinAPRS project is written with what is called CONDITIONAL COMPILE flags. This means that a specific section of source code may or may-not get compiled, depending on what flags are set. We have Macintosh Flags, Windows Flags, and several other internal flags. The objective of the system is to have as much of the code to be common, i.e. compiled in ALL cases, and as little as possible to be specialized code, i.e. compiled ONLY for Mac, or ONLY for Windows. By doing this, we have a much easier system to maintain, and a much more compatible system across different platforms

X-APRS, APRS for X-Windows (UNIX)

At the Dayton Hamfest in May, we had a SUN workstation running a very preliminary version of X-APRS (X-Windows is the Graphical User Interface for UNIX computers). This too is being done with the conditional compiles described above. Doing the development this way allows us to use code that has been around a long time that has been fully tested, thus speeding up development time. We hope to have X-APRS out sometime next year. (1997)

FUTURE

APRS, MacAPRS, WinAPRS and X-APRS are continuing to evolve. These programs have proven themselves to be useful in many more applications than originally imagined. This type of system is a system that takes full advantage of the technology available only in portable radio communications and cannot be replaced with the Internet.

References

[1] ***Automatic AX.25 Position and Status Reporting***
Bob Bruninga, WB4APR
American Radio Relay League, 11th Computer Networking Conference
Teaneck, New Jersey, November 7, 1992

[2] ***Up Front In QST***, December 1994, p 14

[3] ***MacAPRS, Automatic Position Reporting System, A Macintosh version of APRS,***
Keith Sproul, WU2Z and Mark Sproul, KB2ICI
American Radio Relay League, 13th Digital Communications Conference, Bloomington, Minnesota, August 19-21, 1994. pp 133-145

[4] ***Advances in APRS Technology***
Keith Sproul, WU2Z and Mark Sproul, KB2ICI
Proceedings of the 1995 TAPR Annual Meeting
St. Louis, MO, March 1995, pp 55-59.

[5] ***Graphical Information Systems and Ham Radio (The Future of A.P.R.S. Technologies)***
Keith Sproul, WU2Z and Mark Sproul, KB2ICI
American Radio Relay League, 14th Digital Communications Conference, Arlington, Texas, September 8-10, 1995. pp 108-117

Internet Resources

Web sites with APRS Information

http://aprs.rutgers.edu/APRS/
http://www.tapr.org/tapr/html/sigs.html

Automatic Radio Direction Finding Using MacAPRS™ & WinAPRS™

Automatic Position Reporting System

Keith Sproul, WU2Z
ksproul@noc.rutgers.edu
http://aprs.rutgers.edu/APRS/

Abstract

Radio Direction Finding has been around for almost as long as radio itself. Doppler-based RDF systems have been around for quite awhile too. In the recent past, people have developed computer interfaces to Doppler-based RDF systems. APRS has the ability to display the RDF information on maps, giving the user a graphical way to view the RDF patterns.

Over the last few years, the call sign databases available on CD-ROM from several companies have become more and more sophisticated. There are also databases of commercial frequencies and locations available.

Most of us involved in Amateur Radio have experienced situations where we need to track down the cause of an unwanted radio signal, i.e. stuck microphone, improperly tuned equipment, or even a jammer.

With all of the available technology, we should be able to develop a system that zeros in on a location and automatically shows us the possible transmitters in the area.

Computerized Radio Direction Finding

Doppler RDF units have been around for many years. Several years ago, people started trying to get the output of these RDF units to feed directly into a computer. One of the early versions of this was simply a method for reading the status of the LEDs on the RDF unit via a computer interface. Later on, these interfaces became more sophisticated. The current RDF units have serial ports that report not only the direction, but also signal strength indicators. The direction vectors are also reported in much higher accuracy resolution.

This year at the Dayton Hamfest, Agrelo Engineering introduced the DFjr. This unit is a complete computerized RDF unit. During the development of this unit, Agrelo worked with the developers of APRS to ensure smooth operation of their unit and the APRS software.

The 'normal' mode of operation of the DFjr is to have it in a car for doing RDF work. However, this unit also can be configured to be hooked up to a TNC so that each time it hears a signal on the frequency it is monitoring, it will transmit the RDF information over Packet, using APRS protocols.

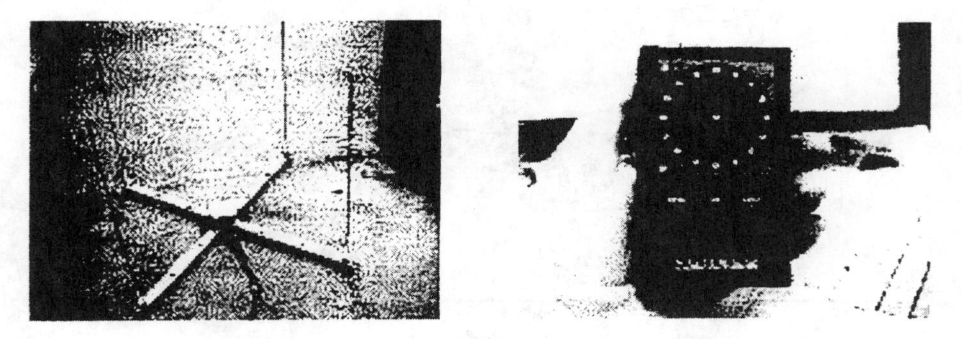

Agrelo DFjr, Computerized Doppler RDF Unit

Computerized RDF and APRS

APRS will take the output of the RDF units and display the information on any of the APRS maps. This gives you a geographical representation of the RDF data. If you have more than one RDF/APRS station participating, then you can get real-time intercept vectors. The first picture below shows WinAPRS and the vectors from a DFjr. The second picture below shows MacAPRS and two stations reporting RDF vectors.

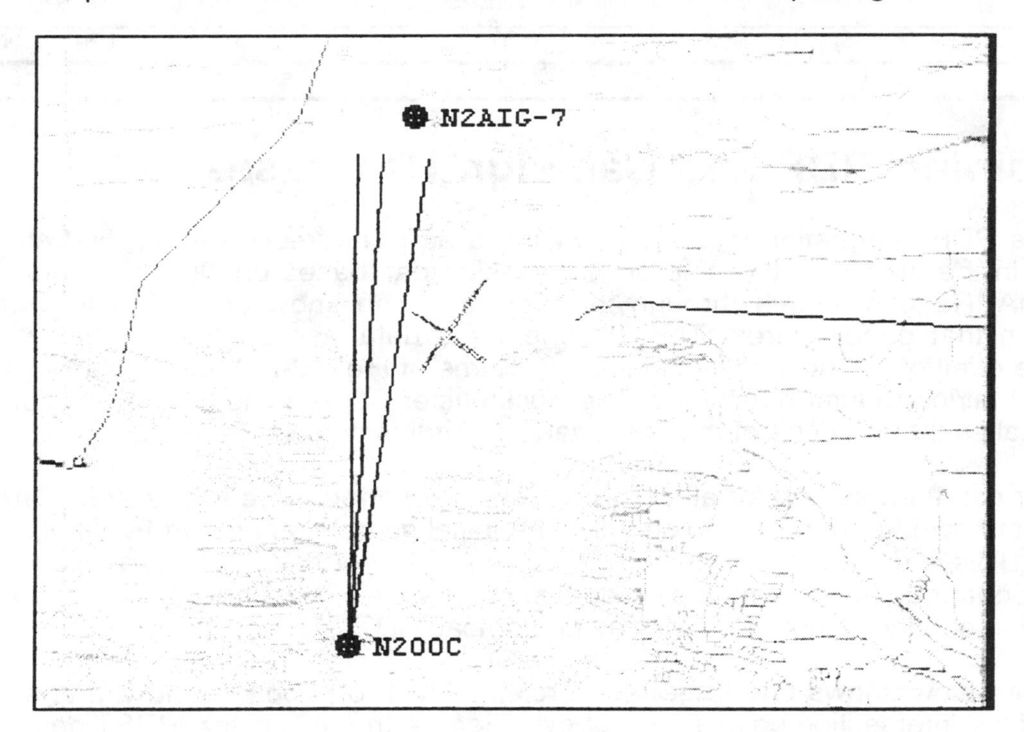

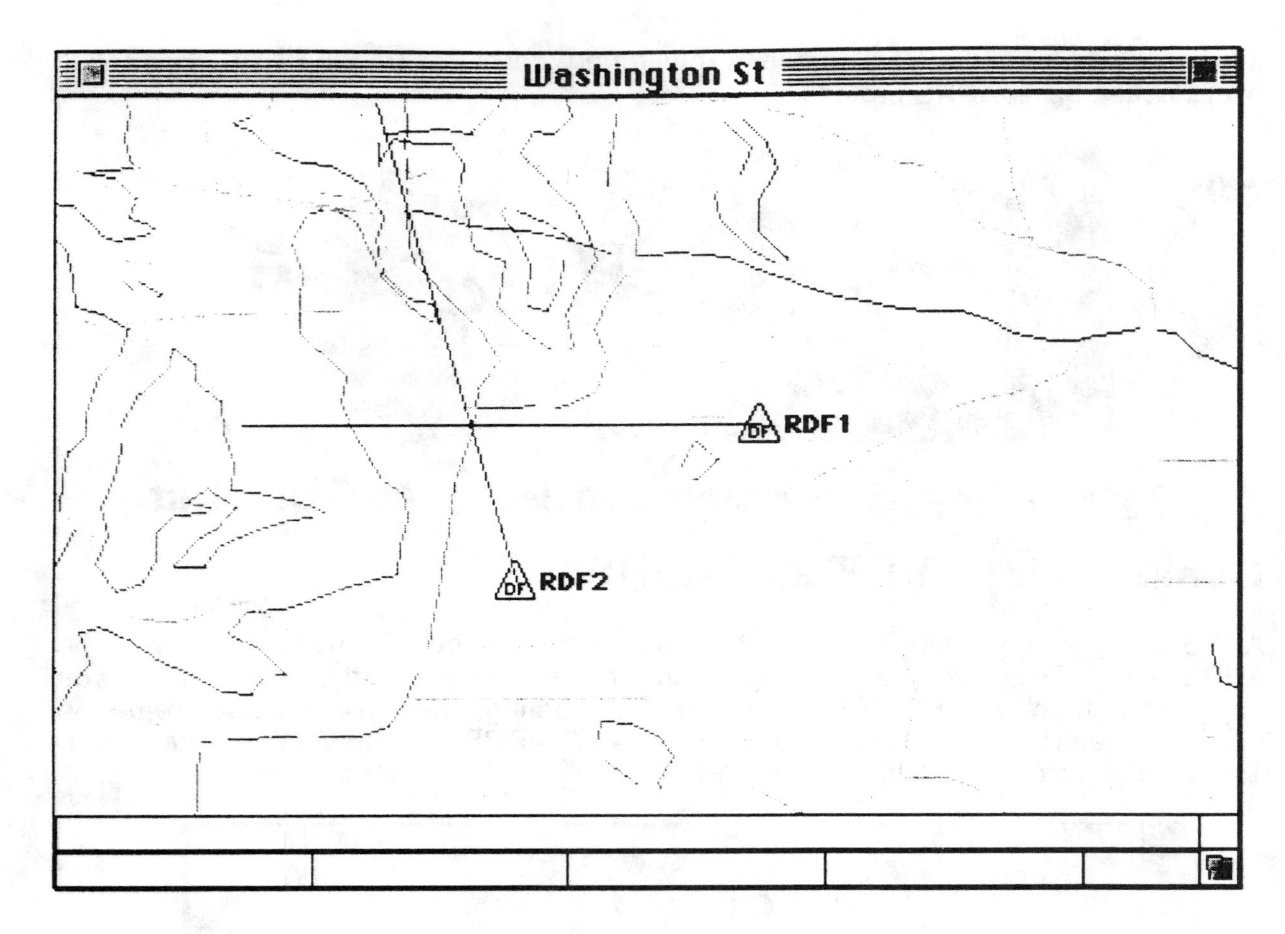

Combining RDF and Call-sign Databases

Once the RDF information lets you know the area of interest, you can find all of the stations in the area with the help of the call-sign databases on CD-ROM. MacAPRS and WinAPRS can search through the database and show you all of the stations located in that general area. This is done via a database containing the latitude / longitude of all of the post offices in the US. Some of the CD-ROMs are starting to add the Zip+4 lat/lon to their databases. The Buckmaster CD was the first to do this. (This, alone, makes their CD one of the best available for this type of use).

The user can then search for all of the call signs reported to be in this area. The user can select how big of an area to search. The initial search is done on the lat/lon of the zipcode. This is done for speed. Then, once this group of data has been selected, it is further enhanced using the Zip+4 data, if available. The chart below shows the information obtained from the Buckmaster Hamcall CD.

The table below shows one page of approximately 110 people found within a 1 mile radius of the intersection point shown above. Realize that this is the FIRST pass based on the 5-digit zipcode. The table shows the actual distance from the intersection of the RDF vectors to each station based on the its zip+4 lat/lon. If the CD-ROM database you are using has the Zip+4 location data, you can double click on each one of the stations in the list and it will show you exactly where that person lives on the map. (Within the accuracy of the Zip+4 system which is generally about 1/2 block).

Selected Call Sign List

Call	LC	First Name	Last Name	Street	City	St	Zip	DOB	Lic Issue	Lic Expire	Area	County	Other	Dist
KB7OAT	N	Jeffrey L	Pullen	4602 S 170th	Seattle	WA	98188-3254	19541017	19910820	20010820	206	King	I	0.2
KB7OHL	P	Eric M	Emry	5565 S 152nd Apt 34	Tukwila	WA	98188-7815	19690708	19930928	20030928	206	King	I	1.3
KB7OTF	P	James L	Quinton	3705 S 172nd	Sea Tac	WA	98188-3628	19490721	19920602	20020602	206	King	I	0.4
KB7RJA	T	Steve P	Olson	3749 S 194th	Seatac	WA	98188-5360	19740715	19921229	20021229	206	King	I	1.5
KB7RYJ	P	Howard T	Mayhew Jr	15325 Sunwood Blvd B402	Tukwila	WA	98188-5726	19410929	19940716	20030119	206	King	I	0.3
KB7SAQ	T	Theresa A	Kennedy	3714 S 152nd 27	Tukwila	WA	98188	19580706	19930119	20030119	206	King	I	0.1
KB7TTM	T	Vaughan F	Philpot	4011 S 152 St	Tukwila	WA	98188-2231	19240913	19930413	20030413	206	King	I	1.1
KB7UFM	T	Zita Joan	Hallstrom	17047 35th Ave S	Seatac	WA	98188-3608	19330820	19930504	20030504	206	King	I	0.4
KB7YTQ	N	Gregory S	Berglund	3754 S 172nd	Seatac	WA	98188-3627	19640501	19930921	20030921	206	King	I	0.3
KC7AHT	G	Nancy B	Schimmelman	645 S Center 260	Seattie	WA	98188	19520215	19940104	20040104	206	King	I	0.1
KC7AVZ	T	Jason E	Parvu	3738 S 164th	Sea Tac	WA	98188-3040	19780212	19940215	20040215	206	King	I	0.5
KC7AZX	T	Margaret K	Thomasson	16432 32nd Ave S	Seatac	WA	98188-3021	19570805	19940222	20040222	206	King	I	0.7
KC7AZY	T	Norma H	Thomasson	16432 32nd St	Seatac	WA	98188	19310214	19940222	20040222	206	King	I	0.1
KC7BNS	T	Danice M	Fisher	17343 Military Rd S	Seatac	WA	98188-3651	19590727	19940329	20040329	206	King	I	0.3
KC7CUS	T	Sidney W	Anderson	3408 S 175th	Seattle	WA	98188-3662	19281125	19940531	20040531	206	King	I	0.5
KC7DBN	P	Douglas W	Hans	16037 45 Th Ave S	Tukwila	WA	98188	19470825	19940705	20040705	206	King	I	0.1
KC7FBP	T	James E	Mitchell	17230 Military Rd S	Seattle	WA	98188-3648	19710318	19950505	20040817	206	King	I	0.2
KC7HPM	T	Todd J	Rogers	3054 S 150th	Seattle	WA	98188-2107	19830509	19941221	20041221	206	King	I	1.4
KC7HVZ	P	Lloyd L	Crabtree	18625 39th Ave S	Seatac	WA	98188-5007	19351112	19950906	20041229	206	King	I	1.1
KC7IGO	T	Diosdado A	Alejo	3511 S 160th St B1	Seattle	WA	98188-2634	19740127	19950117	20050117	206	King	I	0.7
KC7IGR	P	Tina M	Patton	17341 32 Ave S A102	Seatac	WA	98188-4436	19581029	19950201	20050117	206	King	I	0.6
KC7IVC	T	Michael S	Ward	3200 S 176th St 408	Seattle	WA	98188-4072	19670319	19950209	20050209	206	King	I	0.7
KC7KLW	T	Binyamin Y	Levine	16801 33rd Ave	Sea Tac	WA	98188-3132	19461105	19950425	20050425	206	King	I	0.5
KC7LDN	T	Diane L	De Meerleer	4024 F S 158th	Seattle	WA	98188	19480202	19950522	20050522	206	King	I	0.1
KC7MFC	T	Quentin W	Rapp	3806 S 179th St	Seattle	WA	98188-4167	19301015	19950714	20050714	206	King	I	0.7
KC7MUG	T	Jana E	Ward	3200 S 176th St 408	Seattle	WA	98188-4072	19701208	19950819	20050819	206	King	I	0.7

Conclusion

This kind of Geographical Information System has many potential uses within the ham-radio community. This type of search is not limited to ham-radio databases only. There are databases available that contain similar information about commercial transmitters. These databases not only include latitude and longitude, but also actual frequencies etc. Over a year ago, when I started doing demonstrations of this type of capability, many people wanted to have it immediately. However, at that time, the computerized RDF units where either done as build-it-yourself kits, or for the most part, were just not available. Now, with the DFjr from Agrelo Engineering, this type of automatic RDF Unit is easily available and affordable. This type of technology will allow us to do semi-automatic Radio Direction Finding for such things as tracking down interference problems etc.

References

[1] ***MacAPRS, Automatic Position Reporting System, A Macintosh version of APRS,***
Keith Sproul, WU2Z and Mark Sproul, KB2ICI
American Radio Relay League, Digital Communications Conference, Bloomington, Minnesota, August 19-21, 1994. pp 133-145

[2] ***Graphical Information Systems and Ham Radio (The Future of A.P.R.S. Technologies)***
Keith Sproul, WU2Z and Mark Sproul, KB2ICI
American Radio Relay League, 14th Digital Communications Conference, Arlington, Texas, September 8-10, 1995. pp 108-117

Internet Resources

<u>Web sites with APRS Information</u>

http://aprs.rutgers.edu/APRS/
http://www.tapr.org/tapr/html/sigs.html

<u>Agrelo DFjr</u>

http://home.navisoft.com/agrelo/ae.htm

PACKET AND INTERNET

James Wagner, PhD - KA7EHK
31677 N. Lake Creek Drive Tangent, OR 97389
(541) 928-7869 wagnerj@proaxis.com http://www.proaxis.com/~wagnerj

ABSTRACT

Debate is one of the interesting aspects of the packet bbs system. One of the recent debate issues is really quite important to all of us. It concerns the question of bbs mail forwarding by methods other than the ham RF network. Whichever side proves to be "right", (and it is possible that both may be right), the answers to this debate will have an impact on all packet users. KEY WORDS: BBS, NETWORK, FORWARDING, INTERNET

INTRODUCTION

A major debate has raged in the packet bulletin board system over the last year or so. This debate concerns the use of alternate forwarding methods for moving packet messages. In particular, the use of (commercial) internet has been a major issue.

The issues in this debate warrant presenting here because they do represent ideas which are having, and will continue to have, major impact on the future of packet radio.

THE ISSUES, SIMPLIFIED

One viewpoint in this debate argues that the use of alternate forwarding methods (telephone, internet, etc) will result in a deteriorating RF network. The logic is that when alternative methods are used, there is no longer pressure to upgrade existing networks, fix broken ones, or maintain the ones we have. The argument continues with the idea that a network which is allowed to deteriorate will not be there when emergencies arise. And, when emergencies arise, it is also likely that portions of the internet infrastructure will fail. The result will be failure of the ham packet system to perform in emergencies.

The other viewpoint argues that the ultimate responsibility of bulletin board operators is to move "mail". If the ham RF infrastructure is not capable of moving that mail, this argument continues, then they have the responsibility to find some method which will allow the mail to be moved. If that method happens to involve the telephone system or internet (ie, "wire line"), then so be it.

Network Quality

There are a number of factors which combine to represent the quality of a packet network. Of course, not all users have the same idea of quality. None the less, there are a number of general things:

a. links need to be reliable
b. unexpected disconnects should not occur
c. reasonable throughput must be available most of the time
d.hardware is physically reliable

When any of these factors gets worse, usually we perceive the quality of the network to be "worse". A network can remain physically the same, but be perceived as deteriorating because it is not able to handle an increasing message load, for example.

Thus, for a network to remain as a high quality one, just keeping the hardware working is not enough. If the network cannot support increased demands placed on it, then it is not doing its job. Unfortunately, many users consider bbs forwarding to be the culprit, rather than the "miner's canary" which warns us of impending difficulty. To such users, its probably just fine that their local bbs forwards over internet because it makes things seem to work better.

It is also an unfortunate human attribute that often, the quiet wheel does not get improved. If bbs sysops use other methods for forwarding messages, then a visible pressure for network improvement goes away.

Mail Movement

For many bbs sysops, mail movement is their entire reason for participating in ham radio. And, for some, at least, ham radio is just another access method for their bbs. When you have bbs sysops of this sort, what technical methods are they going to choose for linking? Certainly, the most familiar ones. If they are more comfortable with wire-line, that is what they will choose.

While "Clover", for example, may do an excellent job of handling messages via HF, it is probable that more bbs sysops are familiar with wire-line modem technology than they are with Clover. So, is it any wonder that non-ham message passing technology is frequently the method of choice?

It is also likely that arguments that "it is not ham radio" will be quite ineffective. It is probable that many of the bbs sysops in this category do not have a big stake in ham radio and that this argument results in a big "so what?"

On June 30, 1996, W0RLI wrote the author: "Yes, there is still a small amount of traffic handled via satellite, HF digital modes, and long haul vhf/uhf links. In fact, all PRESENTED traffic is easily handled. However, very little traffic is presented to the radio network; it is instead moved via commercial networks. When I first started speaking out on this issue, 18 months ago, about 50% of the long haul traffic was still being carried by radio. That percentage has now reached 0."

The practice of wire-line forwarding is actually having a far bigger impact than it might seem. When strategically located bbs, in widely separated locations, forward almost instantaneously to each other, bulletins arrive in the other area more rapidly and get distributed to those stations which do use RF forwarding. Since the messages are already there when attempted by traditional means, they are rejected. This "capture" phenomenon results in an artifically forced reduction of RF forwarding.

CONCLUSIONS

It seems probable that both sides are "correct". The sad part is that the cases of network deterioration seem to be growing more numerous with the use of forwarding over wire-line. It is also likely, however, that the cause-and-effect is not so clear. Bbs forwarding is moving off the RF network because it is deteriorating in many places and the deterioration is accelerating because there is less reason to keep it up. In other words, it is likely that these two effects go hand-in-hand and neither is the cause of the other.

What does seem fairly clear is that bbs sysops who move their forwarding off the RF network are not doing hams much of a favor. This is, in fact, one area where users can apply pressure, encouragement, and support to sysops. Hams with solid HF experience can help a sysop to set up a reliable forwarding system using Clover, Pactor, Amtor, or packet. Hams with good VHF/UHF network experience can help to make the bbs VHF/UHF packet equipment as good as it can be.

Likewise, bbs sysops AND users can apply pressure and offer assistance to their local ham clubs and packet organizations, and to node operators. Make sure that network capability improvement is planned, that groups involved in packet networking get together and figure out what is needed on a regional basis, and make sure that there is a solid commitment to carrying through on those plans.

Failing all else, bbs sysops in some areas of the country are rerouting messages to avoid forwarding them to bulletin boards which use wire-line forwarding. This is certainly a drastic measure but it is one of the few ways available to avoid the capture effect previously described. As unpleasant as this measure may be, the health of our network(s) may depend on it!

BIOGRAPHY

James Wagner is an electrical engineer & programmer employed by Kalatel Engineering in Corvallis, OR. His work is in the area of embedded controllers and design of system components for the closed-circuit video security industry. He has a BA in Physics from Oregon State Univ, an MS in Electrical Engineering, also from O.S.U, and a PhD in Electrical Engineering from Colorado State University. He has been employed by Tektronix and by the College of Oceanography of Oregon State University. His interest in ham radio began in the 1950's but did not actually get a license until 1979. He has been the advisor to the Oregon State University ARC and the node-op of their packet node. He is the author of "The Amateur Packet Radio Handbook".

From the Universität Kaiserslauten Web Page

SUNSAT: A Micro Satellite Under Construction In South Africa

SUNSAT is a 60 kg, 45 by 45 by 62 cm micro-satellite being designed, built and tested by 22 M.Eng. students at the Electronic Systems Laboratory in the Department of Electrical and Electronic Engineering at Stellenbosch University.

Detailed design started in January 1992, led by Computer and Control System lecturers. SUNSAT was originally designed for a sun-synchronous-type orbit on the Ariane 4 Helios mission, which is ideal for the main imaging payload. However, when launch costs became prohibitive, alternatives were sought.

NASA scientists have learned a great deal about the earth by detailed studies of the magnetic field and the gravitational field, and have arranged for the Danish Oersted microsatellite to be launched as a secondary payload on a USAF Delta II from Vandenberg Air Force Base on the P91-1 Argos mission in January 1996. NASA and Stellenbosch have now agreed to carry SUNSAT into the same orbit instead of a counterweight for Oersted.

In exchange for the launch, Sunsat will carry a precision GPS receiver and a set of Laser retro-reflectors. These will enable NASA to study fine orbital perturbations for gravity field recovery, and for cross verification of GPS and NASA's laser tracking network.The orbit will be the same as Oersted, namely polar, 400 by 840 km. The equatorial crossing will initially be approximately 15:00 UTC, and drift an hour earlier every seventy days.

SUNSAT is a complex microsatellite. Its developers expect they will not have time and manpower to utilize all its possibilities, and hope that other amateurs and universities will become interested in using it once it is fully commissioned. Since this is their first satellite, they recognize this may take many months to get right.

The Amateur Radio communications payload comprises a packet radio service, a 2 m band "parrot" speech transponder, and a Mode-S transponder.

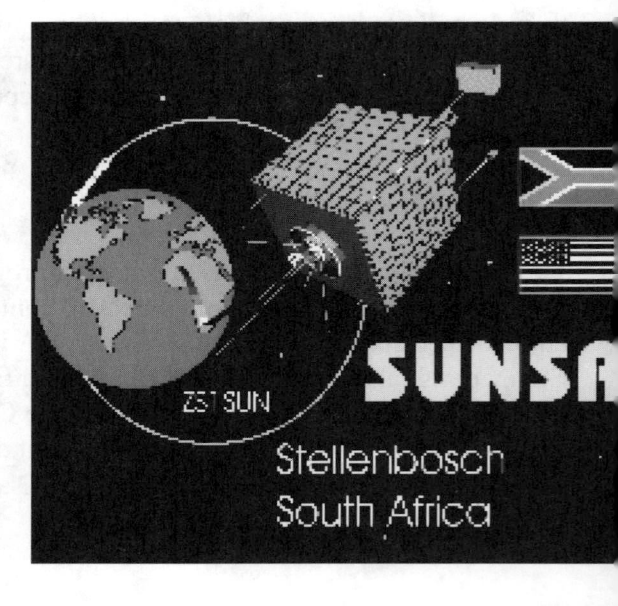

The use of an imaging system necessitates attitude stabilization. Coarse attitude stabilization will be by a gravity gradient boom and by mag netorqueing and is improved by small reaction wheels during imaging. Continuous spacecraft attitude sensing is provided by magnetometers, sun sensors, visible wavelength horizon sensors, and a star sensor provide 1 miliradian accuracy when imaging from the sun-synchronous orbit. The average power of 25 W enables images of South Africa and elsewhere to be taken on a daily basis for real time downlinking.

Availability of excellent linear Silicon CCD sensors able to operate in the visual and near-IR band led to a 3-color sensor system with bands similar to SPOT 4 and LANDSAT 6. These permit "biomass" production monitoring, which is of continuing interest in a "water-short" country like South Africa, for example. A linear CCD sensor with 3456 pixels of 10.7 microns spacing was also chosen. The optical assembly is mounted in a tube which can be rotated forward or rearward for stereo images.

The communications payload provides duplicated synthesized transmitters and receivers for the 2-m and 70-cm Amateur Radio bands and nearby frequencies.

A 23-cm receiver will operate as a fast uplink, or be coupled to the S-band downlink transmitter to provide a straight-through transponder.

The high resolution data will be transmitted in real time via the S-Band downlink to reception stations at Stellenbosch and Johannesburg. Small-area images stored in the RAM disk can be down-linked at much lower rates. For example, a 40 kbyte image covering a 4 km x 4 km area can be downloaded at 9600 baud in about 100 seconds. The SUNSAT team plans to be able to supply such images on request to amateurs once the satellite is fully operational.

At 800 km altitude, a 5 deg elevation footprint has a diameter of 5,080 km and which spans 45 deg in longitude. Radio range varies from 800 to 2,800 km compared to the geostationary range of 36,000 km. Data communication with 10 watts or lower powered transmitters and a dipole antennas is practical, permitting data interchange with low cost terrestrial transceivers. Since large quantities of data can be stored in the satellite, global data transfer is possible. AX25 data protocols will be used to ensure error-free operation.

The 5 watt EIRP S-band downlink will produce a 14.4 dB S/N ratio in a 40 MHz bandwidth at 2000 km slant range for a 4.5 m diameter parabolic dish that has a 100 degree Kelvin receiving station which is current what is being planned for Stellenbosch. By adding an L-band receiver and appropriate switching, a transponder capable of 1 MByte/s with 2 m diameter ground stations can be implemented. Application of the system for Amateur Radio gateway service is possible.

The Amateur Radio payload definition was approved at the SA-AMSAT Spacecon 91 Conference. Store and forward digital packet radio will be provided, including 1200 baud AFSK for compatibility with terrestrial equipment common in South Africa. To provide sufficient uplink channels, one of the 2 m band receivers has four IF sections displaced in 25 kHz steps, and connected to 1200 baud modems. Three 9600 baud modems compatible with the G3RUH standard are carried, and can be switched to various receivers and transmitters. Both the 2 m up/down and 2 m up/70 cm down options will be included, together with full bulletin board facilities. The AMSAT Pacsat Standard Protocols can be supported.

The 2-m and 70-cm downlinks can be switched to 10 watts output, producing a 0.5 µV signal (50 ohm) at 435 MHz and 1.5 µV signal at 145 MHz with 0 dBi receive antenna at full range. This power level will be used over critical areas to provide signal to noise ratios approaching 15 dB for easy reception. At other times the power will be reduced.

A 2 m "parrot" mode repeater is intended especially for Novice category users (under the age of 16). Up-linked speech will be digitally stored and re-transmitted on the same frequency. Novice school users will thus hear the re-transmission and know that they are getting through. The need to learn and apply operating protocols will definitely be experienced!

Stay tuned to the AMSAT News Service (ANS) bulletins for more information about SUNSAT.

By Donald Cox, AA3EK

ACARS: Packet for Airplanes

A new digital communication mode for commercial airliners and business jets

Do you remember the flight attendant on your last airline flight reading off the list of connecting gate numbers as you prepared to land? That information was most likely passed to the flight crew using a new digital communication link called *ACARS*—Aircraft Communications and Reporting System. As complex as it may sound, it's similar in many respects to amateur packet radio.

Until a decade ago, almost all radio links between the ground and commercial aircraft used voice communication to relay weather, position, aircraft performance and departure/arrival reports. The airlines supported the creation of an organization, Aeronautical Radio Inc (ARINC), to run a network of ground stations with which to communicate with airliners anywhere in the world. VHF, HF and satellite links were used to stay in touch with aircraft.

The expansion in air traffic, reduction in the size of flight crews, and the automation of aircraft cockpit and control systems generated the need for a faster and more efficient system for handling communication with aircraft aloft. The result was the creation of ACARS, a digital data link system designed to use existing ground station and aircraft radio equipment, and to enhance air-ground-air communication.

The Heart of ACARS

On board the aircraft, the heart of the ACARS system is a computer that receives data from the cockpit or other aircraft systems and passes it to the VHF radio for transmission to the ground. In the cockpit, the pilots have their own terminal and keyboard to display incoming messages and to compose outgoing requests and reports. A printer provides the crew with a hard copy of the messages.

The ACARS computer can also be linked to other on-board avionics systems. When tied to these systems, the ACARS system can automatically pass aircraft and engine performance data to the ground. Many airlines and aerospace firms routinely collect performance data that is passed to their engineering staffs for analysis and early identification of problems or performance improvements.

ACARS messages currently carry weather reports, arrival and departure time reports and aircraft system data dumps. Plans call for ACARS-derived systems to be used eventually to handle most of the air traffic control messages. Instead of a controller on the

ACARS Hardware, Software and Resources

Decoders and Demodulators

A number of commercial demodulators are available to receive VHF-band ACARS messages, including several under $99. All use the audio output of any VHF receiver/scanner capable of covering the 129 to 132-MHz AM aircraft bands.

AEA ACARS, produced by Advanced Electronics Applications, is an IBM PC-compatible serial-port decoder and DOS-based software system. A 386 or higher CPU is recommended. A software-only version of the system is also available to use with AEA PK-900 or DSP-232 multi-mode controllers, or AEA FAX decoders. The software includes options to suppress the display of messages received with parity errors, write messages to disk, print them, change screen colors, and review and manage log files. A 132-page manual documents ACARS message formats and the extensive reference information necessary for understanding the messages.

The AEA ACARS serial decoder and software package is available from many AEA product distributors, or from AEA, PO Box C2160, 2006 196th St SW, Lynnwood, WA 98036; tel 206-774-5554. AEA ACARS software for PK-900, DSP-232 or AEA FAX product owners is also available separately.

Lowe Airmaster 2.0 is distributed by Lowe Electronics and uses the same basic DOS software as the AEA system, but a different serial-port decoder. Like the AEA software, message fields are broken apart and displayed separately and there are message printing and logging options. It comes with a 24-page manual.

Universal Radio's M-400 is a stand-alone hardware reader that decodes ACARS messages and displays them in the raw, transmitted format. Their M-1200 PC card and M-8000 hardware systems also receive ACARS and display it in the same manner. Universal's product line also includes the ACT-1 PC serial port decoder and DOS-based software system. It comes with a 33-page manual and a copy of the book *Understanding ACARS*. The ACT-1 software includes several message displays and filtering options and displays message fields individually. Universal Radio M-400, the M-1200 PC card, ACT-1, and the M-8000 are available from Universal Radio, 6830 Americana Parkway, Reynoldsburg, OH; tel 800-431-3939.

Books

Understanding ACARS by Ed Flynn, third edition, published in 1995 by Universal Radio. 92 pages.

Complete technical details on the ACARS message format and system architecture is documented in a series of ARINC technical publications, available for $65 to $80 from ARINC Document Section M/S 5-123, 2551 Riva Rd, Annapolis, MD 21401-7465; tel 410-266-4117; fax 410-266-2047.

ARINC Characteristic 724B-2: Aircraft Communications Addressing and Reporting System (ACARS), November 1993. A follow-up to ARINC Characteristic 597-5 covering second-generation ACARS units.

ARINC Characteristic 620-2: Data Link Ground System Standard and Interface Specification, December 1994. Complete technical reference on ACARS message formats.

ARINC Characteristic 635: HF Data Link Protocols, August 1990. Technical reference on HF ACARS system design, message and data link formats.

Cyberspace

Web sites with ACARS information include **http://barbie.epsilon.nl/~bart**, **http://web.inter.nl.net/hcc/Hans.Wildschut**, **http://www.u-net.com/~morfis/acars.htm** and **http://www.arinc.com**.

Table 1
ACARS Message Types

Label	*Message Type*
:;	Data transceiver autotune
5D	ATIS request
5P	Temporary suspension of ACARS
5R	Aircrew initiated position report
5U	Weather request
5Y	Aircrew revision of previous ETA or diversion report
5Z	Airline designated downlink
7A	Aircrew initiated engine data/takeoff thrust report
7B	Aircrew entered miscellaneous message
10-49	User-defined functions
51	Ground GMT request/GMT update
54	Aircrew initiated voice contact request/voice go-ahead
57	Alternate aircrew initiated position report
80-89	Aircrew addressed downlinks
C1	Printer message
F3	Dedicated transceiver advisory
H1	Optional auxiliary terminal message
Q0	Link test
Q1	Departure/arrival report
Q2	ETA report
Q3	Clock update advisory
Q4	Voice circuit busy
Q5	Unable to deliver uplinked message
Q6	Voice to ACARS channel changeover
Q7	Delay message
QA	Out/fuel report
QB	Off report
QC	On report
QD	In/fuel/destination report
QE	Out/fuel/destination report
QF	Off/destination report
QG	Out/return in report
QH	Out report
QK	Landing report
QL	Arrival report
QM	Arrival information report
QN	Diversion report
RA	Command/response uplink
RB	Command/response downlink

3C01 POSWX 1722/18

ground talking to the flight crew via voice radio, the controller will transmit a digital message to the flight crew to tell them to climb or turn. Most new commercial aircraft are equipped with ACARS systems, as well as a significant number of private business jets. Airlines are already experimenting with the use of in-flight satellite communication links and HF links to carry ACARS messages. This would allow global coverage and would reduce the need for an extensive network of ground stations.

You Can Receive ACARS

Most commercial receivers and scanners capable of receiving the VHF aircraft-band AM transmissions can receive the 129 to 132-MHz ACARS frequencies. A growing number of modern Amateur Radio 2-meter transceivers can also cover this band. ACARS frequencies in the US include 131.550, 130.025, and 129.125 MHz; in Europe, 131.725 and 131.525; in Asia, 131.450; and Air Canada has used 131.475 MHz. The 131.550-MHz frequency is considered the primary channel in the US. ACARS-like transmissions on HF have been observed on 6.646, 10.027 and 13.339 MHz on flights traveling the North Atlantic routes.

VHF ACARS reception is easiest for most hams. All you need is an ACARS demodulator connected to the speaker output of your scanner or transceiver and your computer. The demodulator translates the ACARS audio tones into data your computer can understand. Specialized software running on your PC displays the ACARS information on your monitor. See the sidebar, "ACARS Hardware, Software and Resources."

Table 2
ACARS Message Examples

Message label: RA GATE ASSIGN UA837 EWRSFO GATE 82 FREQ 129.5 EON 1844 APU OFF	Translation: United flight 837 from Newark to San Francisco will arrive at gate 82 at 1844 GMT. The aircraft is to turn off its auxiliary power unit upon arrival and contact United operations on129.5 MHz.
Message label: C1 Flight ID: DDAA 2 Message content:- 10012 FROM ^D39 13 13 AN N41063/GL IAD- WX RADAR SHOWS TSTMS DVLPG ALNG THE SOUTH SHORE OF L.I. ALSO ONE LG CELL TO FL600 OVE HUO MOVG SE-20KTS.	Translation: weather advisory message from ground to American Airlines aircraft N41063 near Washington, DC, advising of thunderstorms near Long Island.
Message label: QF Flight ID: CO1854 Message content:- DTW1646EWR	Translation: Departure message from Continental flight 1854, which departed Detroit at 1646 GMT for Newark.
Message label: 40 Flight ID: 0023 Message content:- ARR ORD ARR M20 BAG IAB DEST GATE DALLAS/FT W H6 DENVER H15 KANSAS CITY K15 LOS ANGELES H1	Translation: Message from ground with list of gates for connecting flights at Chicago. The flight will arrive at gate M20.
Message label: 80 Block ID: 7 Msg. no: 1553 Flight ID: DL1722 Message content: KDFW/KLGA .N922D /POS OTT /OVR 2214/ALT 370/FOB 0132/SAT 61 /WND 246025/MCH 746/TRB SMOOTH / SKY CLEAR	Translation: Delta flight 1722 from Dallas to NY is over Nottingham, MD (OTT) at 2214 GMT, flying at 37,000 feet, with 13,200 lb of fuel on board, air temp of –61 deg, wind from 246 deg at 25 kt, speed Mach 0.746 with a smooth ride and clear sky.
Message label: 40 Flight ID: 0023 B Message content: HX TO ORD 28JUL/1745Z POSN-07B KWRD-SEAT ARM-6847007H TRAY TABLE WILL NOT RETRACT/	Translation: aircraft advising ground maintenance that the tray table in seat 7B is broken.
Message label: 5Z Block ID: 4 Msg. no: 1846 Flight ID: US0065 Message content:- /ENG/0938/350/248/740/ 432/M19/M44/199/ 199/860/850/454/433/ 855/880/2750/2700	Translation: a data dump from USAir flight 65 showing it at 93,800 lb of fuel, 35,000 feet altitude, 248 kt speed, Mach 0.74, true airspeed 432 kt, indicated & true outside temp of –19 and –44 degrees C. Pairs of data for the plane's two engines follow: EPR of 1.99/1.99, N1 = 86.0/85.0%, EGT = 454/433 deg C, N2 = 85.5/88.0%, fuel flow = 2750/2700 lb/hr.

Depending on where you live, a simple omnidirectional antenna such as a ground plane or **J**-pole is perfectly adequate. An outdoor antenna is best, but even an indoor antenna can bring you plenty of action. Some have also reported success using only telescoping whips and rubber-duck antennas, but they're the options of *last* resort.

You can expect to receive ACARS transmissions from high-

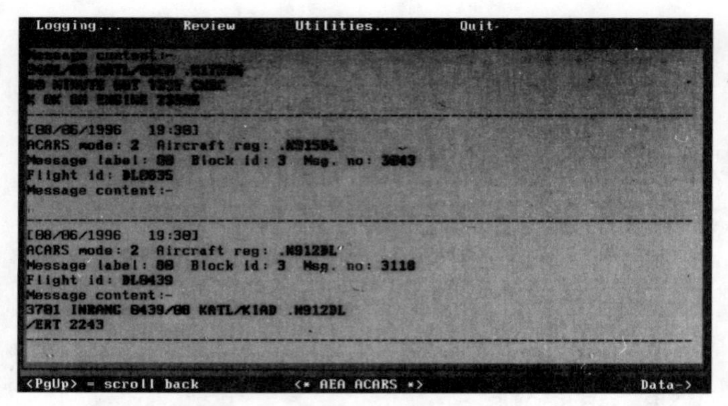

AEA ACARS software decodes received data and displays the results on your monitor.

```
Seq.  Date      Time     M  ADDR    ML B  MSN  FID     MESSAGE
0016 11/08/95 18:16:59 2 ..N791N Q0 6 1650 PI1097
0017 11/08/95 18:17:06 2 .N779AU 52 7 1656 US1041 /IR BWI
0018 11/08/95 18:17:11 J QQq791N Q0 I N)50 P)N79w ~W~
0019 11/08/95 18:17:11 2 .N904DE Q0 J NHLC ;3N0LH |\0
0020 11/08/95 18:17:40 R QQQq 1U _^ 7 Mm1> 0,0P/q |
0021 11/08/95 18:17:50 2 ...NSN)    G  M7A  ,?00Q |P(  PA T(J - 5 H Em   zC
0022 11/08/95 18:18:00 R Q1I5J*>    5 18:0 U-0BnO |   RD   ~)c_UR
0023 11/08/95 18:18:11 J QqJJyUA _^ K N7Oh UARFN/
0024 11/08/95 18:18:11 r QQ-N31U mK 0 2rY> 0,POOA WM~ >0q4P 2N(7KZu7J,US -PQT
                                                   >,x__) P_?,TC  Q>Wx~** ~?
0025 11/08/95 18:18:19 2 ...NS U _^ q :4P> 0,000N
0026 11/08/95 18:18:26 2 .N653UA H1 2 D005 UA0946 #DFB/PIREPUA.E22C246530946KIA
                                                   DEHAM11 0952310CL  122 DATA N
                                                   OT AVAILABLE   38.9713 -77.46
                                                   3223 9  1450  0.2206 22   39.
                                                   1175 -77.50512312  6402-
0027 11/08/95 18:18:32 2 .N779AU 52 8 1903 US1041 /IR BWI
0028 11/08/95 18:18:33 2 .N653UA H1 3 D006 UA0946 #DFB12.0296 30   39.1102 -77.
                                                   16662316 11517-19.0296 31
0029 11/08/95 18:18:35 2 ...N31U _^ 3 -L-  0P0000 |2h
```

The *ACT-1* ACARS software from Universal Radio displays the data in a column format.

altitude aircraft 150 to 200 miles away. If you're lucky enough to live near an airport, you might be able to receive ACARS ground-to-air uplink messages.

Understanding ACARS Messages

The VHF ACARS frequencies carry a large volume of traffic between aircraft and ground stations. The types of information in the transmissions varies widely. It can range from simple arrival or departure reports, to lengthy downlinks of navigation, engine and performance data. Messages might include weather observations and forecasts; departure clearances and checklists; flight plans; navigation positions; aircraft and engine performance data; arrival, departure and delay reports; equipment malfunction reports; crew reports; and connecting gate lists.

Like amateur packet, ACARS transmissions are very short, burst-type signals that last only a fraction of a second. Each message is broken up into a number of subfields. The seven-character address field contains the registration number of the aircraft. The registration number is the official number assigned to the aircraft by the government. International agreements dictate a letter prefix indicating the country of origin. For US aircraft, all registration numbers begin with "N." Shortwave radio listeners and ham radio operators will notice the similarity between these letters and the radio call sign allocations for the same countries. The following give some examples of registration numbers for airliners from several countries: N14245, USA; C-FDSN, Canada; G-BNLR, Britain; D-AIBE, Germany; JA8097, Japan; HB-IGC, Switzerland; and PH-BFH, The Netherlands.

A number of references and computer databases have compilations of registration numbers for aircraft around the world. These can be used to quickly identify the type of aircraft sending an ACARS message, the aircraft operator and other related information.

A series of two-character message labels have been defined to designate the type of message being sent. These include a number of fixed-format messages with key information like arrival/departure times (so-called on/off times) and fuel loaded, as well as a number of labels with variable formats that can be defined by the aircraft operators. These labels are crucial to understanding the type of information in the message; they are frequently the entire extent of the message themselves. For example, if the message label is "51," the aircraft is requesting the ground to update its on-board clock. A short list of these labels is shown in Table 1.

Several examples of ACARS communications between the ground and aircraft are shown in Table 2. ACARS messages use a shorthand style of abbreviations and a good reference is necessary to fully understand the message contents.

Give It a Try!

ACARS monitoring is a fun and relatively inexpensive addition to your ham activities. By following the message traffic, you'll gain an educational insight into airline operations, how flights are planned and flown, the types of daily challenges they experience, and much more. Best of all, you'll quickly appreciate the complexity and sophistication of today's airlines and air traffic control systems.

PO Box 11130
Washington, DC 20008-0330
e-mail 74507.3446@compuserve.com

SCS PTC-II Multimode Controller with PACTOR-II

By Steve Ford, WB8IMY
Managing Editor

It's astonishing to think about how far we've come in HF digital communications in such a short time. Prior to 1982, Baudot RTTY was the norm. It was a great mode for casual conversation, but it lacked the means to detect errors. If the RTTY signal you were copying fell victim to noise, fading or interference, the result was gibberish on your screen. (Before the PC revolution, the "screen" was often teletype paper!)

AMTOR burst (pun intended) onto the scene in '82. It gave us text that was almost error-free, and filled the bands with those odd *chirp-chirp* sounds. A few years later HF packet came into its own. Although far from ideal in terms of dealing with the vagaries of HF, packet at least gave us the ability to exchange the entire ASCII character set, as well as binary files.

Special Communications Systems (SCS) created PACTOR, starting with PACTOR-I, which hit the airwaves in 1991 with sophisticated data compression and the remarkable ability to "reconstruct" corrupted data. Reaction from the ham community was swift and positive. It wasn't long before PACTOR-I became a staple in virtually all multimode controllers.

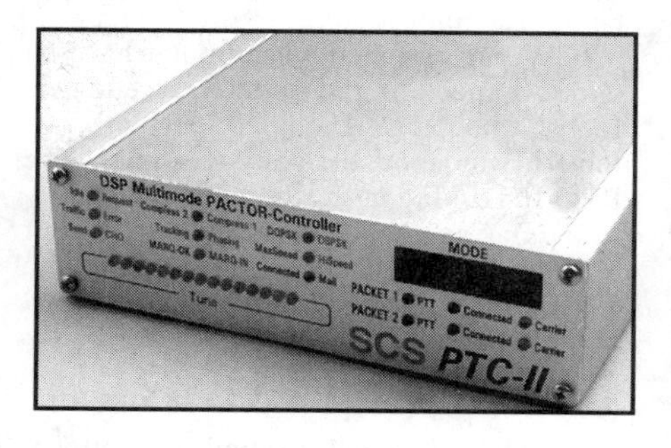

BOTTOM LINE

A top-of-the-line digital box at a top-of-the-line price. This one's for *serious* digital enthusiasts.

Just when it seemed that the HF digital waters had calmed a bit, CLOVER and G-TOR appeared. Both modes offered superior performance, but hams seemed content with PACTOR-I and were reluctant to make a change. Even so, CLOVER and G-TOR found dedicated, if somewhat limited, audiences.

As we progress toward the end of the 20th century, we have no fewer than six digital modes competing for the hearts and minds of HF-active amateurs. If you listen to the HF digital subbands for any length of time, you'll discover that while the venerable RTTY is still quite popular for digital contesting and DXing, PACTOR-I reigns as the undisputed king of the "burst modes."

In the current "Tower of Babel" digital climate, will hams accept a *seventh* mode? If they're offered an HF digital mode that promises the ultimate in performance, will they flock to it in numbers sufficient to dethrone PACTOR-I? That's what the folks at SCS—the creators of PACTOR-II—are betting on.

PACTOR-II

I once heard a ham describe PACTOR-II as being PACTOR-I on steroids. There may be some truth to this analogy! PACTOR-II retains many of the features that made PACTOR-I so popular: Support of the compete ASCII character set; memory ARQ to reduce the number of retransmissions; on-line data compression to speed transfers; automatic data-rate selection; and independence from sideband selection (it doesn't matter if you operate with your rig set to LSB or USB).

What PACTOR-II brings to the party is considerable. Using differential phase-shift keying (DPSK) and innovative convolutional coding (with a true Viterbi decoder), PACTOR-II boasts data transfer rates at least *three times* faster than PACTOR-I. Under optimum band conditions, the rate could soar to six times faster. In terms of effective data rates, that translates to about 800 bits/s. PACTOR-II has the option of using Huffman coding for data compression (the same as PACTOR-I), or Pseudo-Markov coding (PMC) for even greater performance and efficiency.

One of the most astonishing aspects of PACTOR-II is that it achieves all this wondrous performance while using only 500 Hz of spectrum (at –50 dB). You need fast hardware and digital signal processing (DSP) techniques to make all of this practical. That's why you won't find PACTOR-II included in the current crop of multimode controllers. In fact, the *only place* you'll find PACTOR-II at the moment is in the PTC-II.

The PTC-II

When I first eyeballed the front panel of the PTC-II, one image came to mind: A Christmas tree. There are no fewer than 30 LEDs populating the panel, along with a dot-matrix LED readout. They're all quite informative, telling you whether the data is flowing without errors, which compression scheme is in use, if you have messages waiting in the mailbox and so on. In practice I found that I didn't have time to focus on any one of them. Only the dot-matrix **MODE** display and the tuning indicator saw much use. The **MODE** display is handy when you're tuning around in the *listen* mode. If I came upon a PACTOR signal, the display would immediately indicate whether it was a PACTOR-I or PACTOR-II conversation.

The tuning indicator is a real gem! It is simply a row of dual-color LEDs with a **TUNE** LED in the center. As you tune across a PACTOR-II signal, for example, the LEDs on the right- and left-hand ends flash red, then green, as you "arrive" on frequency. At that point you make slight adjustments of your VFO until the **TUNE** LED is centered. With a little practice I found that I could tune in a PACTOR-II signal in a matter of a few seconds.

On the back panel are two packet radio ports. (Yes, the PTC-II supports *simul-taneous* HF/VHF communication—up to *three* channels at once!) You can't operate packet with the standard PTC-II, though. If you want packet capability, you must purchase additional modules from SCS. A 300 baud modem is not yet available, but SCS offers an AFSK modem for 1200 and 2400 baud, plus an FSK modem for 4800, 9600, 19,200 and higher baud rates. The labeling of the HF ports can be a little confusing. There is a port labeled **AUDIO** and another labeled **CONTROL**. The **AUDIO** port provides the audio inputs and outputs, as well as the push-to-talk lines. **CONTROL** refers to the interface between the PTC-II and your transceiver's computer-control port, if it has one. For basic operating you need only construct a cable for the **AUDIO** port, although you can do some very clever things with the **CONROL** port, as we'll discuss later.

Communicating with your computer is as easy as running an RS-232 cable between the PTC-II and your computer's serial port. The PTC-II package includes a freeware program called *PlusTerm* on a 3.5-inch diskette. I tried to load *PlusTerm* using its installation batch file and had a devil of a time. I eventually gave up and worked around the problem.

To my dismay I discovered that *PlusTerm* is strictly a DOS program. As a guy who's been spoiled by *Windows*, I found it painful to use. The good news is that the terminal program already available in *Windows* works very well with the PTC-II. The even better news is that Wilfried Max, DL1XAM, has developed a full-featured *Windows* program specifically for the PTC-II. (You can get a demo version by sending a formatted 3.5-inch diskette with $10 *cash* (no checks) to Wilfried Max, DL1XAM, Lisbeth Bruhn Str 18, D-21035 Hamburg, Germany. For more information see the SCS site on the Web at **http://www.scs-ptc.com/software.html**.)

The only serious weak spot in the PTC-II package is the manual. The slender volume does an excellent job of explaining PACTOR-II, but gives poor coverage to other aspects of the PTC-II. For example, I didn't know that the PTC-II could send and receive CW until I discovered this feature by accident while reading through the list of commands. Some of the unit's most interesting features are described vaguely, leaving users very much on their own.

On the Air

After hearing and reading so much hype about PACTOR-II, I was dying to give it a try. My first problem was *finding* a PACTOR-II station. With the PTC-II in the *listen* mode, I scanned through the 20-meter digital subband, but all of the "burst" links were using PACTOR-I. I wasn't entirely surprised. There are only a handful of PACTOR-II stations in the US and a limited number in Europe.

After about 30 minutes of hunting I finally copied a link with the ON4OB mailbox. As I tuned in the signal, the **MODE** indicator on the PTC-II flashed **PT-2**. Bingo! When the station signed off, I sent a connect request to the mailbox. Two seconds later we were connected and the data stream was flowing nicely. The band was in poor shape, and conditions were worsening by the minute. I requested a list of his station equipment just as his signal took a deep fade. To my astonishment, the list began marching across my monitor despite the fact that I could *no longer hear his signal*!

Later the same afternoon I witnessed a dramatic demonstration of PACTOR-II's ability to withstand interference. I had connected to FG5FU and was downloading information when a RTTY station fired up right on top of us. The link seemed totally unaffected by the RTTY interference. In fact, the text continued to zip by faster than I could read!

Of course, the PTC-II features its own mailbox, with a 512-kB message capacity (expandable to 2 MB with the optional static RAM extension). It offers some versatile remote-control functions as well. The mailbox accepts connections in PACTOR-I or II, or AMTOR.

It is important to point out that PACTOR-II, as implemented in the PTC-II, is entirely "backward compatible" with PACTOR-I. That is, you can work PACTOR-I stations and they can work you. When you connect to a PACTOR-II station, the exchange begins in PACTOR-I and then automatically "upshifts" to PACTOR-II.

AMTOR conversations were even harder to find than PACTOR-II. I spent the better part of an afternoon searching for stations running AMTOR and came up empty. That evening I finally encountered a few on 80 meters. Once again the PTC-II did an outstanding job of maintaining links in the face of considerable noise and interference. After using the PTC-II on both PACTOR modes, however, AMTOR seemed almost primitive.

The PTC-II had a surprise in store on RTTY. The DSP filter performance that proved itself so well on the other modes did a *superb* job with RTTY. I encountered a number of stations using RTTY on 20 meters and enjoyed several live conversations (all of my PACTOR-II connections were with mailbox systems). The PTC-II provided wonderful copy even when signals sank into the noise. This unit might be a great contest box!

CW performance was about what you would expect. If the code was sent perfectly, the PTC-II copied perfectly. If the fist was poor, the copy was poor. The PTC-II's tuning indicator made CW reception quick and easy. You simply tune your rig until the LEDs begin "bouncing" from left to right in sync with the signal.

Fascinating Extras

The PTC-II includes a number of bells and whistles that you won't find on other controllers.

I had to chuckle when I read that the PTC-II can function as a *DSP audio filter*. No kidding! The unit provides an audio output line (on the **CONTROL** port) that you feed to a small amplifier and speaker. (The audio from your rig is already feeding into the HF transceiver port.) At your keyboard you enter the AUDIO menu and select either an adjustable-bandwidth CW filter, a "denoiser" or a notch filter. I noticed dramatic performance with the CW filter; the skirts seemed incredibly sharp. The denoiser worked best on CW, while giving a somewhat "hollow" sound to SSB signals. The automatic seek-and-destroy notcher did a fine job of eliminating "tuner-uppers."

And what about that **CONTROL** port? If your transceiver provides computer control, you're in luck. The PTC-II can use it to do one or both of the following:

- Control the frequency of your radio from your computer keyboard, or remotely from another station.
- Set up an automatic scan on frequencies you select in advance. You can enter up to 16 frequency "channels" and command the PTC-II to scan through all of them. As it does so, it commands your radio to jump from one frequency to another. For example, you can configure the PTC-II to scan on 16 different frequencies from 80 meters through 10 meters. If your friend wants to connect to your PTC-II mailbox, he has 16 different frequencies he can try, depending on band conditions.

No special software is required for rig control. The PTC-II comes preconfigured to work with Yaesu, Kenwood, or ICOM rigs. You simply "tell" the PTC-II which brand you own.

DPSK requires exceptional frequency stability. The PTC-II includes automatic offset and drift compensation using a tracking algorithm that follows the received signal.

Although the manual only mentions it briefly, the PTC-II has the ability to vary your output power *automatically* during PACTOR-II contacts. If conditions are good, the PTC-II will reduce the audio level applied to your transceiver, thereby reducing your output. If conditions go to pot, audio levels automatically increase to push your radio up to full power. The default setting for this command is OFF, so you must switch it on yourself.

Speaking of transmit audio, some transceivers offer audio line-input ports on their rear panels (these are often labeled **AFSK**). If the PTC-II doesn't seem to provide enough audio through this port to drive your rig to full output, no problem! Just use the **PSKA** command at your keyboard to adjust the transmit audio level.

Finally, the DSP architecture of the PTC-II provides extraordinary flexibility for the future. SCS plans to offer free firmware updates that will allow the PTC-II to operate on FAX and SSTV. All you'd have to do is place the software in your computer and implement the PTC-II's **UPDATE** command.

Is PACTOR-II for You?

If you demand high-performance HF digital communication regardless of cost, PACTOR-II may be your dream come true. There is no question that PACTOR-II offers superb performance, and its implementation in the PTC-II is outstanding. Is PACTOR-II better than the other contenders for the high-performance market—CLOVER or G-TOR? PACTOR-II proponents would answer "yes," but CLOVER and G-TOR disciples would argue otherwise, depending on specific conditions. Monitor the Internet e-mail "reflectors" and you'll be treated to an ongoing battle among those who champion these modes.

In the meantime, the average ham has yet to embrace PACTOR-II, CLOVER or G-TOR in significant numbers. They seem comfortable with PACTOR-I, especially since it provides excellent performance in controllers costing less than $300. As one fellow commented on the air during one of my tests "PACTOR-I already gives me the ability to swap text at rates faster than I can read or type. Isn't that enough?" Perhaps, but the benefits of PACTOR-II go beyond speed. The robust nature of PACTOR-II makes global communication possible for low-power stations, or stations with less-than-optimal antennas, *or both*.

Regardless of its technical merits, the fate of PACTOR-II is likely to be decided in the marketplace. If PACTOR-II is to gain wide acceptance, hams will have to be convinced that the benefits are worth the substantial investment.

Manufacturer: Special Communication Systems GmbH, Roentgenstrasse 36, D-6345[illegible] Hanau, Germany; tel 49-6181-23368; e-mail **info@scs-ptc.com**; WWW **http://www.scs-ptc.com**. Manufacturer's sug- gested retail price, $950; SCS RCU remote control amplifier unit, $195. (PacComm Packet Radio Systems Inc, Tampa, Florida, manufactures a nearly identical version of the PTC-II in the US, under a license from SCS, for $995. Call 800-486-7388 or e-mail **ptc@ paccomm. com**.)

Packet Perspective

Stan Horzepa, WA1LOU*

Geminids Packet Meteor-Scatter Test Results

The Geminids packet meteor-scatter test was conducted on December 14, 1995. This was the second test resulting from my proposals to expand the horizons of packet—so to speak—by trying something truly *different*. Compared to ordinary packet operating, you can't get more different than bouncing your signals off the fiery trails of meteors!

The first test was conducted during the peak of the Perseids meteor shower in August 1995. Participation was limited then because the test was planned on the spur of the moment. Word was spread via the Internet, but not in print. Those of you anticipating a test announcement in Packet Perspective had to wait until the second test was announced in December.

With the greater advance publicity, I hoped for an improved turnout for the December test. Mother Nature played havoc with our plans, however. The turnout was less than expected, and the results were not as encouraging as the August totals. (Weak-signal meteor-scatter operators also report a rather mediocre Geminids event.)

On 6 meters, the best meteor-scatter propagation was achieved by Dave Gaytko, WD4KPD, in North Carolina, who received packets from Ed Myszka, KE4ROC, in Alabama, 545 miles away. KE4ROC was also heard by Joe Fadden, N9QLW, in Chicago, a 491-mile path. For sheer duration, it's hard to beat the *59* packets that WD4KPD received from WA4NTF over a 445-mile path.

The results on 2 meters were more impressive. The longest-distance report came from Jim Mollica, N2NRD, in New Jersey, who copied packets from Tom Cash, K4ZQX, in Chattanooga, 648 miles away. Syd Chiswell, W2ICZ, in Buffalo also received K4ZQX's packets, a 619-mile path.

There was some confusion when stations received what they thought were meteor-scattered packets, only to find out that the so-called DX was coming from APRS HF gateways within their tropospheric earshot. For example, many stations received this or a similar packet from KFØZH:

KFØZH>APRS,GATE*,WIDE/V:@141518z4447.06N/09329.48W_202/000/T025/R000/P000/U-II

If you missed the asterisk next to GATE, you thought you hit meteor-scatter pay dirt. However, those who spotted the asterisk realized that the packet was relayed by an HF gateway station, not by a shooting star.

Soapbox

I saw several short burns at 2145Z during the second and third quarter-minute transmit periods, but not fast enough for the software to decode. Otherwise, only locals and no monitored packets during the 24-hour period (WE8W). Much to my dismay, I discovered that our 2-meter preamp failed several hours into the Geminids test. I wonder if anyone received our beacons? (WB8IMY, op at W1AW). I ran all night from FM03 on 50.620 MHz and didn't hear a thing beaming at 225°. I suspect the snow and freezing rain didn't help (VA3MW). 145.79 MHz is totally crowded here in the Philly-to-New-York area. Had a blast! I suggest readers grab *MSSOF42F.ZIP* from **ftp.funet.fi** in the **pub/ham/vhf-work** directory. Excellent program for meteor scatter work. The AZ/EL beam heading charts alone are worth the download! (N2NRD). Would like to do some more of this. I have a 2400-baud AFSK modem that may be interesting to try, but I think that everyone who is serious with this mode needs to convert to PSK modems (WD4KPD). Nothing, nada, zip. A 40-foot dish with 25 kW ERP pointed from Maryland toward Chicago showed no receive hits. OSCAR array pointed at Atlanta and Florida showed no hits (WB4APR, op at W3ADO). Sorry to report that I heard nothing this time. I didn't have my beam ready in time. Will try again (KB4LCI). Thanks for your efforts getting this going. I don't do a lot of VHF work and this has been my only experience with meteor scatter. It's fun! The ice storm did a number on my antenna (SWR of 5:1!), so I saw nothing overnight (AAØSM, op at WDØGNK).

December 14, 1995, Geminids Packet Meteor-Scatter Test

6 meters (50.62 MHz)

Stations Reporting			*Stations Received*				
Call Sign	*Grid*	*Ant*	*Time*	*Call Sign*	*Grid*	*DX*	*Pkts*
N2HMM	FN21se	H 300	0				
NO2G	FN21	V	0				
WD4KPD	FM15ml	H 235		KE4ROC	EM64rg	545	2
WD4KPD	FM15ml	H 235		WA4NTF	EM81	445	59
N9QLW			1406	KE4ROC	EM64rg	491	1
VA3MW	FM03	H 225	0				

2 meters (145.79 MHz)

Stations Reporting			*Stations Received*				
Call Sign	*Grid*	*Ant*	*Time*	*Call Sign*	*Grid*	*DX*	*Pkts*
KB1TX		V	0				
W1AW	FN31jt	H 225	0				
WA1LOU	FN31mp	H 225	0				
KA2JAK		V	0				
N2NRD	FM29lq	H 270	0457	K4ZQX	EM75ic	648	1
W2ICZ	FN02xa		1242	K4ZQX	EM75ic	619	2
WB2NGZ			0				
W3ADO		H	0				
KB4LCI	FM08xi	V	0				
KD4LCR			0				
W5SR	DM95cf	H 315	0				
KE6AFE			0				
KC7CO		V	0				
AA8SF		V	0				
KA8JMW		H	0				
WE8W	EN85		2145 undecodable short burns received				
WDØGNK	EN34rb	H 135	0				

Notes

V-vertical antenna
H-horizontal antenna; the following number indicates beam heading
0-no packets were received
DX-distance in miles between the two stations
Pkts-number of packets detected

*One Glen Ave
Wolcott, CT 06716-1442
e-mail: **stanzepa@nai.net**

On-air Measurements of HF Data Throughput
Results and Reflections

Ken Wickwire (KB1JY)
kwick@mitre.org
KB1JY@WA1PHY

1. Introduction

For the past year a team of colleagues and I[1] have been collecting and analyzing data on the throughput and other characteristics of various ARQ protocols available to hams and commercial users for HF work. This activity was motivated by discussions (especially among hams) about the relative merits of the new HF digital modes, such as PacTOR, GTOR, CLOVER II, CLOVER-2000 and PacTOR II. Since the discussions often centered on throughput in various conditions, and we were already running several of the protocols, we decided to see for ourselves. This paper describes our assessment approach and measurement campaign, gives a summary of our main conclusions, and lists some findings worth noting before protocol choices are made and protocol performance is compared. The paper treats the packet and TOR modes in detail. More extensive reports on CLOVER II, NOS TCP/IP and the ALE orderwire will appear elsewhere.

2. Our Approach to Throughput Measurement

The randomly varying HF "channel"; that is, the combination of propagation conditions (fading, dominant ionospheric layer, etc.), and propagated and local noise, is generally agreed to be the worst radio channel. Over the past 20 years, powerful DSP techniques have been developed to tame this wild conduit and put it to work for data transmission, even when it resists being used for voice traffic. These techniques are now embodied in a surprisingly large and growing number of data transmission protocols whose performance is often impressive by HF standards. What people mean when they say (or write in an advertisement) that one of these protocols is better than another is not always clear, however.[2]

HF data transmission protocols can be divided for general discussion into two categories: those with automatic repeat request (ARQ) and those without. In some cases, the latter are called forward error correction (FEC) protocols, because they use FEC but not ARQ to control errors. ARQ protocols, which are almost always combined with FEC in modern systems, generally deliver error-free data, although there is no guarantee that the data will be delivered quickly. Since many users (especially military and governmental users, and operators of forwarding stations) demand error-free transmission, ARQ protocols have come to dominate technical discussions of late. For ARQ protocols, the definition of throughput, for example, is relatively straightforward; for protocols without ARQ, which can deliver erroneous data, the concept of HF throughput is more difficult to define. For these reasons we have decided to concentrate on ARQ protocols in our assessments.

There are three basic ways to assess the throughput (and most other kinds of performance) of an HF data transmission protocol. First, you can try to devise a

---

[1] See the Acknowledgments and reference list at the end of the paper.

[2] Although we recognize this shortcoming, we sometimes suffer from it ourselves.

mathematical model of what happens during data transmission with it, convert the model, if necessary, into computer code, and run the code (or your pencil) to assess (i.e., predict) performance. While this is sometimes advocated as the best approach by those too lazy, poor or otherwise constrained to try other approaches, it frequently produces unconvincing or incomprehensible results. Many believe that the modeling approach is best suited to the design stage of a new protocol rather than to the performance assessment stage.

Second, you can connect a pair of working systems through an analog or digital HF channel simulator, which can be set up to accurately produce various levels of some of the main phenomena of the HF channel, like multipath spread, fading and noise (usually Gaussian). Using a channel simulator with real hardware and software produces statistically repeatable results and allows reasonable—if not necessarily convincing—comparisons of different systems operating in so-called "standard channels," namely the ones whose statistics are programmed into the channel simulator. A channel simulator cannot, however, reproduce the statistical variations in transmission quality that occur on a real HF channel; it can't faithfully reproduce those caused by non-Gaussian (e.g., impulse) noise, intermittent and random interference by man-made signals with various waveforms, day-night transitions, and polar and equatorial propagation anomalies.

The third approach is through on-air measurements. This has the advantage that any one measurement is in a sense completely realistic and convincing, but the disadvantage that the conditions in which the measurement was taken are not generally repeatable. This means that producing *statistically* convincing assessments with this approach requires that a large number of measurements be made (resulting in a large sample-size) and that attention be paid to realistic and representative path lengths, power levels, antennas, diurnal variations and the spreads (variances) of performance statistics. This takes time and a lot of cooperation from several outlying stations.

We believe that a combination of channel simulator and on-air measurements leads to the most convincing assessment of ARQ performance in the HF bands.[3] The simulator creates repeatable channel extremes, while properly conducted on-air measurements comprise channel conditions the simulator hasn't been (or can't be) set for. This paper discusses a measurement campaign we've pursued in that belief for the past year. We should note that although our results allow an informative comparison of the throughputs of the protocols we've treated, the past year's measurements need to be continued to cover all seasons with all protocols and a wider range of sunspot numbers.

For our on-air measurements we try to write software that allows tests to be run automatically, so that the mistakes that we all tend to make during manual time-recording and data-entry can be avoided. (Sometimes—as with CLOVER and NOS TCP/IP implementations,—protocols come with their own interface software, and we use the existing software capabilities. That leads to some manual data logging.) So far, our software has been written in C for the Macintosh operating system, but it would work (with different I/O calls) on different operating systems.

With our software, we always measure file transfer time from the start of character-by-character uploading of a file to the sending modem to the time that a "message saved" (or equivalent) sent by the receiving station arrives via the sending modem to the test

---

[3]Of course, this is only true when the two approaches produce results that agree, at least qualitatively. For some simulator results that agree qualitatively with our measurements, see Refs. 1, 2 and 3.

program[4]. Waiting for a "message saved" makes the transfer times a little longer than they would be if a human operator (or program) at the receiving station recorded the error-free arrival of the file. That in turn makes the throughputs slightly lower (i.e., more conservative) than they would be otherwise, but that is a small and reasonable price to pay for measurements that require only one program and no operator intervention during tests.

In the next section we'll describe the philosophy behind our choice of equipment and file types for most of the measurements made during our campaign.

3. Our Philosophy for Choosing Equipment to Use and Files to Be Sent

In choosing equipment (e.g., the KAM for PacTOR assessment) for our tests, we have been motivated not only by expediency (we have KAMs), but by the view that assessments of most interest to the most (prospective) operators are those of "common" operating setups; that is, ones widely available at competitive prices, and ones that offer a wide choice of operating modes and good technical support. While it is no doubt true that some implementations of PacTOR, for example, may have higher throughput than others because they use A/D quantization of bit energy or more advanced filtering, they are probably not in wide enough use to be part of a "common" operating setup. Nevertheless, if we had the time and money to buy and test all possible pairings of implementations of a particular protocol, we would gladly do it, since performance of the "best" or the "official" implementation is obviously of interest. In the meantime, we unselfishly invite others to fill in the gaps left by our work.

Likewise, we have chosen at this stage ASCII English text files of various sizes to compare the transfer capabilities of protocols. With due respect to the many who probably send text files written in other languages, we believe that sending such files represents a "common" application of the HF ARQ protocols described below. It should be borne in mind that languages other than English and German, and files with a non-standard distribution of characters (e.g., all upper-case characters), may benefit very little from the Huffman text compression used in current PacTOR implementations.

When a protocol like PacTOR, GTOR or CLOVER II comes with defaults for some of its "protocol-tuning" parameters (e.g., GTTRIES and GTUP for GTOR and BIAS for CLOVER II), we have used these defaults. This has been based on the belief that a common setup would not have these parameters changed. (Optimal tuning of such parameters is an area that should be looked at, however, and a few operators have recently started to do so.) For packet, on the other hand, we consider good values of PACLEN and MAXFRAMES to be highly dependent on channel conditions, and we juggled these values frequently to increase throughput in our tests (see below).

Finally, our philosophy says that if a protocol or common implementation offers data compression, then it should be used (if there's a choice) unless we think it might seriously expand a file (see the section below on data compression). This means that in the case of PacTOR we used Huffman compression and in the case of CLOVER II, we used the "PKLIB" compression (probably a Liv-Zempel-Welch variant) offered by the standard (i.e., "common") P38 terminal software provided by HAL with the modem.

---

[4]That is, we don't include linking and "negotiation" times in our throughput calculations. Others may view these times as legitimate components of transfer times.

In the next section we'll discuss some of the protocols we have assessed in the past year of our campaign along with some more advanced ones we may get to later.

4. ARQ Protocols Developed for HF Use and Their Throughput

Table 1 at the end of the paper lists most of the ARQ protocols that are in common use on the air[5]. With the exception of the last three, all are used in amateur work, and the last three, developed originally for military use, will probably enter the amateur world in some form in the next few years.

The table classifies each protocol according to its modulation scheme, signaling bandwidth, forward error correction capability, ARQ scheme, channel rate, character format, compression capability and measured throughput for "standard" (mainly lower-case) English text files. The throughputs with a ◊-symbol beside them have been measured as part of our campaign, with enough samples for statistical significance in current sunspot conditions. The other throughputs are based on channel simulator measurements. The measured throughputs for packet and the TOR modes are from an aggregate of short near-vertical-incidence skywave (NVIS) and longer one-hop skywave (OHS) paths. For CLOVER II and the ALE engineering orderwire, the throughputs are from only NVIS paths. (We expect to begin OHS tests with CLOVER II this summer, and to publish ALE, NOS TCP/IP and CLOVER results this year.)

It should be kept in mind that in agreement with our measurement philosophy, for our packet and TOR throughput measurements we have used Kantronics KAMs with firmware version 7.1 or higher. Other PacTOR implementations than the KAM's may yield higher or lower throughputs than ours. Note also that we have used the HAL P38 for all of our CLOVER II measurements; more expensive models, like the PCI-4000, have the computing power to select a 16-symbol signaling set, and may produce higher throughputs.

5. Differences Between NVIS and OHS Throughput for TOR and Packet

NVIS throughput is generally lower than throughput over "standard" one-hop skywave (OHS) paths; that is, fairly long paths on which fading (and resulting inter-symbol interference) is relatively slight, and average signal-to-noise ratios are comparatively high. In fact, one-hop skywave measurements paint a relatively optimistic picture of what operators can expect in day-to-day communications over HF.

However, some protocols appear to improve more than others when you go from NVIS to OHS operations. Tables 2 and 3 below (reprinted from recent papers listed in the References) give NVIS and OHS throughput and other statistics for AMTOR, PacTOR, GTOR and packet. (Recall that for packet, we juggled PACLEN and MAXFRAMES to increase throughput.)

Throughputs in the tables are in characters/sec and times are in seconds. The first column gives the average throughput and its standard deviation, the average throughput per Hertz, the standard deviation of the mean throughput and the maximum observed throughput. The second column gives the number of links and the mean and standard deviation of the

[5] A recent newsgroup FAQ on signalling formats lists a number of ARQ protocols in use in Europe, the CIS and Asia that we never heard of, so we may be misleading our readers with this statement. Most of these protocols may be rather old and inefficient, like AMTOR, but we can't be sure.

"link time." The third column gives the number of "negotiation times" and the mean and standard deviation of the negotiation time. Link time is the time (in seconds) between sending the link command and receipt by the program of the "LINKED TO" notification. Negotiation time is the time between sending the link command and the start of message-file transfer. In most cases there are fewer negotiation than link times because we started measuring the former part way through the campaign. The fourth and fifth columns give the means and standard deviations of the transfer time and the number of transferred characters.

The standard deviation of the mean (equal to the standard deviation of the throughput divided by the square root of the sample size) is an assessment of the variability of the mean itself (which has its own statistical variability). The `sd_mean`'s in the tables suggest that our sample sizes are big enough to give us pretty high confidence that if we collected many more throughput measurements under roughly the same conditions, we would not get average throughputs that differed from the ones above by more than about a character per second.

To calculate the average throughputs per Hertz [`E(tput/Hz)`], we divided the average throughput by the average signaling bandwidth. We calculated the latter using the formula for "necessary telegraphy bandwidth" (from the 1992 Dept. of Commerce *RF Management Handbook*) BW = *baud rate* + 1.2 x *shift*, where *shift* for most of our TOR and packet tests was 200 Hz. For AMTOR, the baud rate is of course 100; for PacTOR, GTOR and packet, we used the rough average of the baud rates chosen automatically in the PacTOR and GTOR modes and manually in packet. Our estimates of these average baud rates were 150 (PacTOR), 200 (GTOR) and 200/300 (NVIS/OHS packet). The resulting average bandwidths were AMTOR: 340 Hz, PacTOR: 390 Hz, GTOR: 440 Hz and NVIS/OHS packet: 440/540 Hz.

The majority of NVIS measurements were at 3.606 MHz LSB, with some at 7.085 MHz LSB and 1.815 MHz LSB. They were made during the winter over all daylight hours and also in the evening, after dark; a few were made in the middle of the night. Interference usually prevented throughput tests from about six to ten in the evening (2300Z-0300Z) on 3.606 and 7.085 MHz.

Most of the OHS measurements were at 10.141 MHz LSB. About 20% were taken at 3.640 MHz, 14.075 MHz, 14.123 MHz or 18.075 MHz, all LSB. These measurements were made during the winter and spring over all daylight hours and also in the evening, after dark. However, interference often prevented throughput tests from about six to ten in the evening (2300Z-0300Z) on 3.640 MHz. The NVIS and OHS tests covered roughly the six-month period from November, 1995, to April, 1996.

All measurements were made using transmitter output of around 100 watts, and all stations generally used sloping longwires or dipoles. These setups can be viewed as embodying average station capabilities. NVIS paths (in New England) were from 30 to 200 miles long and OHS paths (on the east coast and from New England to the midwest) were from 400 to 1200 miles long.

In discussing the TOR and packet results let's start with some observations on NVIS and OHS communications quality in general. First of all, note that we haven't collected data on the fraction of tries in each mode that we were successful in linking, "negotiating" and transferring a file. However, we have found that over OHS paths, the three TOR modes and packet can get files through in the absence of strong interference on most tries during the day. This is in contrast with our NVIS results, which showed that except during the mid-morning and mid-afternoon "windows," packet and AMTOR had transfer probabilities well below one.

Under difficult conditions (especially those on NVIS paths leading to marginal SNRs) PacTOR occasionally out-performed GTOR in terms of throughput, although GTOR has higher average throughput. This seems to confirm the rumor that GTOR needs high SNRs for high performance. However, this "role-reversal" happened much less frequently over OHS than over NVIS paths.

In the early evening on both NVIS and OHS paths, there was sometimes increased interference on the frequencies we used. During these periods of interference it was rare to see a file transferred. (An automatic link establishment (ALE) system, such as prescribed in MIL-STD-188-141A, could probably have found a frequency without interference.)

Table 2. Statistical Summary of NVIS Throughput Data

Mode	E(thruput) sd(thruput) E(tput/Hz) sd_mn(tput) max_tput	No_links E(lnk_tm) sd(l_tm)	No_neg_tms E(neg_tm) sd(neg_tm)	E(xfer_tm) sd(xfer_tm)	E(No_char) sd(No_chr)
AMTOR	5.20 cps 1.13 cps 0.015 cps/Hz 0.08 cps 6.33 cps	226 3.02 s 3.16 s	70 82.4 s 30.1 s	473.5 s 234.0 s	2358.1 974.7
PacTOR	17.83 cps 5.50 cps 0.046 cps/Hz 0.30 cps 25.10 cps	344 5.44 s 8.39 s	95 38.7 s 22.7 s	146.1 s 90.0 s	2452.7 1110.1
GTOR	23.52 cps 10.06 cps 0.053 cps/Hz 0.55 cps 44.12 cps	335 5.54 s 10.30 s	76 58.7 s 30.9 s	120.0 s 95.8 s	2531.7 1580.3
packet	5.68 cps 3.53 cps 0.014 cps/Hz 0.25 cps 17.34 cps	197 8.73 s 10.48 s	119 102.7 s 66.9 s	556.7 s 367.6 s	2484.9 1043.1

Turning to particulars, you can see that AMTOR and PacTOR average throughputs don't differ much on OHS and NVIS paths, although there is a slight tendency toward greater statistical variation (as measured by standard deviations) on NVIS paths. This similarity of average throughputs may explain why you don't hear much about differences between performance on long and short paths in these two modes.

The big story is the differences between GTOR and packet performance on long and short paths. Average GTOR throughput on OHS paths was almost 50% higher on OHS paths than on NVIS ones (32 char/s vs 23 char/s). This may reflect the presence of consistently higher signal-to-noise ratios (SNRs) on OHS paths, since GTOR is said to thrive on high SNRs and suffer more than the other modes on low ones.

Packet throughput was two-and-a-half times higher on long paths than on NVIS ones (16 char/s vs 6 char/s). Some of this difference may have been caused by the fact that we restricted all our NVIS tests with packet to 200-baud operation. Although we based this restriction on observations of performance, it's possible that a more aggressive choice of baud rate on packet during the mid-morning and mid-afternoon "NVIS windows" could have raised NVIS packet throughput somewhat. However, this does not explain all of the improved performance, whose source must be the better OHS channel (fewer packet bit errors).

Another striking difference appeared in the average packet negotiation times (OHS: 35 s, NVIS: 103 s). (Recall that negotiation time is the difference between the time a connection request is sent and the time file transfer starts.) This difference in average negotiation times apparently reflects the fact that the negotiation process for a packet BBS upload, which involves transmission of frames of various sizes, exposes packets at 200 baud much higher bit-error rates on NVIS paths than 300-baud negotiations over OHS paths.

Table 3. Statistical Summary of OHS Throughput Data

Mode	E(thruput) sd(thruput) E(tput/Hz) sd_mn(tput) max_tput	No_links E(lnk_tm) sd(l_tm)	No_neg_tms E(neg_tm) sd(neg_tm)	E(xfer_tm) sd(xfer_tm)	E(No_char) sd(No_chr)
AMTOR	5.70 cps 0.80 cps 0.017 cps/Hz 0.08 cps 6.33 cps	104 2.62 s 3.81 s	92 69.7 s 15.2 s	543.2 s 109.8 s	3009.6 98.1
PacTOR	20.19 cps 5.49 cps 0.052 cps/Hz 0.44 cps 25.00 cps	153 4.70 s 6.53 s	139 40.8 s 29.4 s	176.1 s 105.2 s	3058.8 308.5
GTOR	32.30 cps 9.88 cps 0.073 cps/Hz 0.79 cps 44.12 cps	158 4.44 s 7.96 s	144 50.9 s 21.7 s	119.9 s 102.6 s	3126.6 501.4
packet	15.67 cps 4.58 cps 0.029 cps/Hz 0.44 cps 24.59 cps	108 6.46 s 8.50 s	108 34.6 s 17.7 s	221.9 s 141.4 s	2975.0 259.8

Maximum observable TOR throughputs were about the same for NVIS and OHS paths, although, as mentioned above, individual measurements came closer to their maxima more often on the long then on the short paths. On packet, we achieved maximum OHS throughput of about 25 char/s vs about 17 char/s for NVIS.

6. Discussion of Packet Results

Our packet experiments over both NVIS and OHS paths have led to surprising results in view of what we have read on newsgroup discussions and elsewhere. For example, over OHS links we have consistently achieved average packet throughputs two to three times higher than average AMTOR throughputs, although not quite as high as PacTOR, and only about half of the GTOR average (see Table 3 above). The parameters we have adjusted to do this are PACLEN, MAXFRAMES, FRACK, SLOTTIME, RESPTIME and PERSIST, and we have done all our OHS file transfers at 300 baud. (Since we have tried to choose frequencies and times where there is little interference, we have set PERSIST very high and FRACK, SLOTTIME and RESPTIME low for aggressive use of the channel.)

These high packet throughputs have been achieved, however, *only during the day*, and by means of very frequent, manual, changes of PACLEN and MAXFRAMES. Furthermore, we have managed to find frequencies that were by and large free of significant interference from other signals (this appears to rule out most of the 20m band). For example, we have often been able to transfer files over both NVIS and OHS paths with combinations like PACLEN = 100 and MAXFRAMES = 5 in the absence of contention, which may be a revelation to some hams who have tried HF packet.

As a general rule, as packet begins to work in the morning on our links, values of PACLEN/MAXFRAMES around 40/1 work best. From mid-morning till late afternoon, combinations like 80-100/4-7 often lead to high throughput. As the bands begin to deteriorate, it's back to near 40/1. PACLENs greater than about 120 bytes usually suffer too many bit errors on our NVIS and OHS links to be worth trying.

After about 5 PM local time during the winter, throughput rapidly falls, and for most of the evening, getting files through in any mode was difficult. (We got some NVIS transfers through in the middle of the night during the winter, but we didn't try any OHS transfers in the middle of the night.) On our links, trying a lower ham frequency in the evening usually led to increased interference, against which none of the modes did a great job.

Our experience with HF packet on OHS and NVIS links has convinced us that an adaptive protocol that adjusted HBAUD, PACLEN and MAXFRAMES using feedback on throughput could go a long way toward polishing HF packet's tarnished reputation. However, with much better systems now available at reasonable prices, it is probably no longer worth developing such a protocol.

7. The Effects on Throughput of Data Compression and File Type

Three of the protocols we have assessed over the air provide one or more types of optional or hard-coded data compression: PacTOR has (optional) Huffman compression, GTOR has hard-coded Huffman and run-length compression and CLOVER II with the HAL interface software has one or more hard-coded compression techniques from the so-called "PKLIB" suite. Other and future protocols may also include one or more compression capabilities[6]. Of course, even when a protocol doesn't include compression, the user is free to compress his files off-line before he sends them, provided that the protocol can handle the compressed format and the receiving station can de-compress the files. (For an introduction to data compression see Ref. 7.) As mentioned above, we almost always choose the Huffman option in PacTOR transfers of English text files.

How much a file gets reduced by a compression technique is strongly affected by the file's type and the technique's approach, so that the user must have some understanding of the interplay of the two if he wants to use compression for high throughput.

In general, the closer the distribution of a file's ASCII characters to the distribution of characters in "typical" English (or other language to which the Huffman code has been tailored) text, the more Huffman will compress the file. The more repeated contiguous characters or bytes (e.g., spaces) in the file, the more run-length coding will compress it. The more repetitions of byte-pairs in a file ("an," "th" in "the," etc.), the more so-called Markov coding (multi-level Huffman) will compress it. The more repeated byte strings in the file, the more "dictionary-based" methods like Lempel-Ziv-Welch (LZW) compression will squeeze it[7]. Finally, the bigger the monochromatic patches (e.g., big expanses of white background) in a graphics file, the smaller a graphics compression technique (like those used in JPEG and for GIF files) will make it.

These facts means that if you send a text file consisting of a high proportion of upper-case characters with PacTOR, you won't get much benefit from Huffman, which relies (in most PacTOR implementations) on a fixed text-character distribution in which certain lower-case characters (like "e") occur with relatively high frequency. Likewise, if you compress a file off-line (e.g., zip it), you produce a compressed (8-bit, or binary) file that looks a lot like a pseudo-random string of bytes. If you then apply a built-in compressor like one from the PKLIB suite used in the P38 CLOVER software from HAL, you will find that the "compressed" file is actually a bit larger than the zip-file. (Of course, this is all right if the zip-ing did a good job.)

On the other hand, if you try to send an uncompressed executable (".exe") image as a binary file with CLOVER II and the HAL software, you'll find that the already pseudo-random structure of most executable (binary) files is likewise expanded rather than compressed by PKLIB. To get efficient transfer by CLOVER II, you should compress executables off-line before submitting them to the HAL P38 terminal software.

Graphics files (not yet the main focus of our throughput experiments) are another story. If they're GIF or JPEG files, they're generally already compressed, so CLOVER and most other compressors won't make them any smaller[8]. PICT and BMP files, on the other hand, are not compressed, and often have big monochromatic chunks, so that the

---

[6]The TACO2 protocol suite developed by the DOD for transmission of battlefield-situation graphics over HF has data compression as an integral part.

[7]For this reason, it is not a good idea to compare throughput for (cooked-up) files that consist of repeated sections (for example, those made by repeatedly pasting a section to the end of the file). LZW will generally compress such a file by much more than 50%, whereas Huffman will only compress it by as much as it compresses the first section. The resulting comparison is therefore probably unfair to the Huffman, since in many cases one would send just the first section of such a file with the advice that it is to be repeated *N* times at the receiver for whatever reason.

[8]The popular shareware EXPRESS terminal program that also runs the P38 and other HAL CLOVER hardware offers built-in compression of files and tailored compression and transmission of graphics images. Since EXPRESS doesn't (yet) fall under our definition of a "common" implementation ("comes with the modem"), we don't cover it here.

PKLIB compressor(s) in the HAL software usually make them a lot smaller, with correspondingly higher throughput.

So far in our NVIS experiments with CLOVER II using both compressed and uncompressed files we have found that compression plays a crucial role in the relatively high average throughput (above 40 characters/s) we report in Table 1. (Recall that this average applies to compressed English text files, and that OHS transfers are not included in our CLOVER data.) The average CLOVER II throughput over NVIS paths for *uncompressed* files is only around 25 char./s, which is about the same as the GTOR throughput of text files[9].

As we pointed out in Section 3, we have not generally sent off-line-compressed files for throughput comparison, so as not to penalize unfairly common implementations of protocols (like AMTOR and standard AX.25 packet) that can't easily handle binary files. The field of throughput comparison using compression techniques that aren't part of "common implementations" is a wide open and important one.

8. Concluding Remarks

One of the conclusions we've reached in our throughput assessments is that hams and others need to separate long from short distance paths when they compare HF throughput performance of ARQ modes, especially the amateur GTOR and packet modes. Some moderation of opinion on HF packet performance may also be called for.

For HF data transfer, data compression plays a crucial role in increasing throughput, and it should always be used when it significantly lowers file or message size and the receiving station is equipped to handle decompression.

We hope that our throughput data will further clarify discussions of the HF digital modes. Our results should put throughput measurements of PacTOR II and other HF data-transmission systems in perspective. We have plans to report someday on the performance of some of those newer modes, and encourage those already in a position to do so to publish their measurements.

Acknowledgments.

I'm grateful to Mike Bernock (KB1PZ), Dennis Gabler (KB5HVN), Doug Hall (KF4KL), Richard Harrison (NT2Z) and Bob Levreault (W1IMM) for handling numerous requests to put their stations on the air in various modes on various frequencies, and for regular mailbox cleanings.

Mike Bernock and Bob Levreault made a number of useful comments on the text of the paper.

References

1. Young, T., et al., "A Preview of HF Packet Radio Modem Protocol Performance," 13th ARRL Digital Communications Conference, Bloomington, Minn. 1994.

[9] As already mentioned, GTOR applies Huffman and run-length compression to all transferred files.

2. Reynolds, P., “HF channel Simulator Tests of Clover,” *QEX*, Dec. 1994.

3. Riley, T., et al., “A Comparison of HF Digital Protocols,” *QST*, July 1996.

4. “HF Throughput Truths,” in Packet Perspectives, *QST,* Feb. 1996.

5. Wickwire, K., et al., “On-air Measurements of HF TOR and Packet Throughput, Part I: Near-Vertical-Incidence-Skywave Paths,” *Digital Journal*, March 1996.

6. Wickwire, K., “On-air Measurements of HF TOR and Packet Throughput, Part II: One-hop Skywave Paths,” *QEX,* June 1996.

7. Nelson, M. and J-L Gailly, *The Data Compression Book*, 2nd Ed., M&T Books 1996.

By Phillip Nichols, KC8DQF

Portable Packets

Turn a tiny "palmtop" computer into a dynamite packet station!

The complete palmtop packet station is smaller than my license plate! From left to right, my HP-200LX palmtop computer, BP-2 modem and HTX-202 transceiver.

I have been interested in computers since the "old days" when I dabbled with a Commodore Vic-20 and my parent's Atari 400. Back when the Macintosh was still an Apple IIe! I kept pace with the changing technology while I upgraded my computer skills. In recent years I've relied on my notebook PC and a Hewlett-Packard HP-200LX "palmtop" computer.

Actually, my notebook PC has taken a few trips to the computer doctor. At one point it was absent for two months! It was during this time that I became more interested in exploring the capabilities of my HP-200LX. I also learned of a fascinating hobby: Amateur Radio.

After four weeks of study and the inevitable exam, I was rewarded with my Technician Plus ticket. A few months later I earned my Advanced. I was eager to see what I could do with this new hobby. Before long, I discovered *packet*.

What's Packet?

In case you're a stranger to the term as hams use it, packet is a method of connecting computer systems and networks by radio—typically using FM transceivers. Many hams use packet to connect to bulletin board systems (BBSs) where they download files and exchange electronic mail (e-mail). Others use packet to help them hunt DX and gather contest points (*DX PacketCluster* networks), while some enjoy tracking moving objects on computer-generated maps (*APRS*—the *Automatic Packet Reporting System*).

Besides the radio and a computer, all you need to get started is a *terminal node controller* (TNC). Think of a TNC as a radio modem and you'll understand how it works. A TNC takes data from your computer and turns it into audio tones that the radio can transmit. A TNC also takes audio tones from the radio and converts them to data for the computer. In addition to the conversion task, TNCs assemble data into the proper packet formats, check for errors in received packets (and request replacements if necessary) and perform many other functions.

TNCs can be stand-alone devices—little boxes that sit next to your radio and computer. But TNCs can also be created in *software* with just a tiny external modem to convert tones to data and vice versa. Software TNCs have the advantage of requiring little external hardware. That makes them ideal for *portable* packet operating!

Creating a Palmtop Portable Station

Portability is important to me. I live with the most restrictive antenna requirements you can imagine: *no* outdoor antennas allowed! Like many "indoor hams," I hang antennas in my living room. (I'm blessed to be married to a forgiving wife!)

I started to wonder if I could use my palmtop for packet. With a palmtop and an FM rig, I could log on from *anywhere*—even from inside a shopping mall or at work.

The key ingredient turned out to be the BP-2 by Tigertronics. It comes with *Baycom 1.4* software and a modem (a small 2-inch-square custom analog/digital modem based on the old Bell-202 standard). The *Baycom* software makes a computer behave like a TNC. In computerese we say it "emulates" a TNC. Baycom is available free on various landline BBSs, CompuServe, and on the World Wide Web.

Palmtop Packet Resources

Tigertronics
PO Box 5210
Grants Pass, OR 97527
tel 800-8BAYPAC, or 541-474-6700
fax 541-474-6703;
World Wide Web **http://www.tigertronics.com**

The Palmtop Paper
(a newsletter devoted to palmtop computing)
Thaddeus Computing
57 E Broadway
Fairfield, IA 52556
tel 800-373-6114

Other palmtop software on the Web at: **ftp://eddie.mit.edu/pub/hp951x/NEW/** includes:

REAL95.ZIP—A satellite-tracking program. Slow, but effective.

GRID.ZIP—A grid-square locator

ARS-LOG.ZIP—A logging program database for the HP-200.

GEO44CGA.ZIP—Graphical display of day/night illumination of the Earth as well as the twilight "gray line." Can also plot contacts by latitude and longitude.

I connected my BP-2 modem to my Radio Shack HTX-202 handheld according to the BP-2 instructions, keeping the cables as short as possible. It worked like a charm!

The Future

I'm planning to operate a land-mobile packet station while I'm taking vacation trips this summer. I'd also like to try *aeronautical mobile* packet from a private airplane just to see how it would work. Of course, a portable packet station like mine is ideal for public-service work. You could, for example, assist disaster relief by relaying emergency traffic. The possibilities are endless!

If you're living in a restricted environment, or if you are intrigued by the idea of portable packet operating, get your hands on a palmtop and give it a try!

e-mail maccabeus@gatecom.com
packet kc8dgf@kb8rmx.amps.org
WWW http://www.gatecom.com/~maccabeus/hampage.html

QST

Software For The BP-2 and BP-2M

From The Tigertronics Web Page

SOFTWARE.DOC 3/24/97

To help you get started with your BP-2M Modem, we have put together this list of compatible shareware and commercial programs.

In this document we will tell you who makes the program, what modes they have to offer, and where you can get your copy! We will tell you how to configure each program for use with your BP-2M and make comments about configuration or operation. At the end of this document you will find a list of incompatible programs and the reasons why each of them will not work with the BP-2M.

Tigertronics in no way endorses or supports any of the programs listed in this document. We have compiled this list only to aid you in locating programs that are known to work with the BayPac MultiMode. As you should expect, some programs will work better than others. Many work better or worse depending on your specific equipment configuration. We recommend that you download and evaluate several programs to determine which works best in your particular application. This list represents only a sampling of the software that is available. We will make an ongoing effort to refine and update this list as we become aware of suitable programs. The latest release of this list and many of the programs mentioned, can be downloaded from our Internet Web Site at: http://www.tigertronics.com

We would appreciate any comments you may have about programs that you are using with the BayPac. We welcome all comments - Pro or Con! We would like to add your comments to the list, for all to share (if appropriate!).

INTRODUCTION

Before we get started, here are a few notes on how this information is put together. The programs in this document are listed in alphabetical order by program name. If the program is from a commercial source then the company's name will be listed. If the program is shareware then you will find the word "Shareware" instead.

To find out where you can get your copy of the program, look for the "Source:", followed by one (or several) numbers. These numbers reference an Internet Web Page, a telephone BBS number, or the authors address (Sources are listed at the end of this document).

The "Comments:" line will give you the information that you need to configure the program, along with any tips and suggestions that might be helpful.

The "Min Requirements:" line tells you what the computer hardware requirements are for each program (as per the programs documentation).

The last thing we need to mention is that the BP-2M has many different "Modes" which allow it to operate with the various programs. The number that you see following the line "BayPac Mode:" indicates which Mode you must select in the BPMODE program to be compatible.

AEA ACARS   Advanced Electronics Applications (AEA)

Source: 1   BayPac Mode: 3, 6

Comments: Outstanding performance. Long range reception of ACARS signals is possible do to the height of aircraft above the earth.

Min Requirements:

Supports: ACARS (Aircraft Communications and Reporting System)

AEA FAX III

Advanced Electronics Applications (AEA)

Source: 1

BayPac Mode: 3, 6

Comments: If you want to use a comm port other than COM1, be sure to specify the comm port when you start the program: FAX /2 for COM2, FAX /3 for COM3, etc.

Min Requirements: IBM compatible XT, CGA, DOS 2.1

Supports: CW, FEC (Sitor-B, Amtor-B & NavTex) 45-100 baud
RTTY (Baudot, ASCII 7 & ASCII 8) 45-100 baud
FAX (288, 352 & 576 IOC) 60, 90, 120, & 240 LPM

BAYCOM

The BayCom Team

Source: 2, 19

BayPac Mode: 1, 2

Comments: BayCom is probably the most popular packet terminal software out there. In our testing, it consistently out performs all other TNC emulating software. BayCom is available as a shareware product or as a commercial product. Version 1.4 is the latest (and the last) SHAREWARE release. BayCom v1.6 is the current commercial version and it offers many new features. In addition to printed and bound manuals, BayCom v1.6 supports comm ports 3 and 4, keyboard macros, and 9600 baud operation (requires 9600 baud modem). Improved security during remote operations, new transfer protocols, and a screen saver are just a few of the enhancements.

Min Requirements: IBM compatible XT or better, EGA or better, COM1 or COM2, DOS 3.2 or later.

Supports: Both Baycom v1.4 and v1.6 support 300 and 1200 baud packet. BayCom v1.6 also supports 9600 baud operation (requires 9600 baud modem). Baycom v1.4 and v1.6 support binary (not the YAPP protocol) and ASCII file transfer. Baycom v1.6 also automatically saves 7Plus encoded files to disk (decoding of the files is NOT done by BayCom). Both versions of BayCom support remote operations, and both can be set up to operate as a mailbox or mini-BBS.

EZSSTV.ZIP

Easy SSTV v3.0 - Shareware

Source: 3, 19, 20, 25

BayPac Mode: 3, 6

Comments: EZSSTV is a reduced-functionality version of the popular Pasokon TV. Version 3.0 NOW TRANSMITS, displays over 16 million colors, and has three additional modes! This program is mouse driven

	and includes a built-in paint program for creating or editing your images. EZSSTV is easy to use, and it offers great Slow Scan performance!
Min Requirements:	IBM Compatible 386DX, 640K, VGA, COM1 or COM2 only
Supports:	SSTV Robot 36, Martin 1, Scottie 1 and Scottie DX Wraase 120 and 180 sec.

FBB515.ZIP	FBB v5.15 - Shareware
Source: 4, 19, 23	BayPac Mode: 1, 2
Comments:	FBB is a popular packet radio BBS and Server program that is identical in operation to the W0RLI or WA7MBL BBS systems. Due to its many options and features, the installation and setup of FBB is more complex than other packet programs.
Min Requirements:	IBM compatible XT, 640K, Mono CGA Uses the TFPCX ax.25 driver (see comments under TFPCX210.EXE).
Supports:	300 and 1200 baud packet. Gateways, chat, and conferencing. YAPP and ASCII file transfer. A telephone interface is also supported for those who want a "landline" connection!

GP161B.ZIP	Graphics Packet v1.61 - Shareware
Source: 5, 19, 24	BayPac Mode: 1, 2
Comments:	Graphics Packet is a mouse-driven packet terminal program. It comes with an ASCII text editor, and a screen saver. A BBS function is also built in, but we were unable to test it due to a lack of English documentation.
Min Requirements:	IBM compatible XT, EGA Uses the TFPCX ax.25 driver (see comments under TFPCX210.EXE).
Supports:	300 and 1200 baud packet. ASCII and 7Plus file transfer.

GSHPC12.ZIP	GSHPC v1.2 - Shareware
Source: 6, 19, 20	BayPac Mode: 3, 6
Comments:	GSHPC offers good SSTV performance, and is easy to use once you become familiar with the program. The program has separate transmit and receive windows, and it supports BMP and TIFF images with up to 16 million colors! A paint program is also built in. Select these options when configuring GSHPC: For PTT control use "Com_x RTS-pin". For demodulator use "Com_x DSR-pin". For modulator use "Com_x TxD-pin".
Min Requirements:	IBM compatible 386DX, 640k, 1MB VGA

Supports: SSTV Martin 1, 2, 3, & 4
Scottie 1, 2, 3, 4, DX
Robot Color 12, 24, 36, & 72
Robot BW 8, 12, 24, & 36s
SC-1 8s, 16s, & 32s
SC-2 30s, 60s, 120s, 180s

HAMCOM31.EXE — HamComm v3.1 - Shareware

Source: 7, 19

BayPac Mode: 3 or 4 for CW, 3 for all other modes, or 6 (rx only)

Comments: This version of HamComm has many new features and improvements. There is now a tuning indicator on the main screen so that you do not have to change screens to make a tuning adjustment! A new PACTOR listen mode has also been added (requires registration). With its great performance, and continuous improvements, its easy to see why HamComm is one of the most popular multimode programs.

Min Requirements: IBM compatible XT, 640k, CGA, DOS 3.0

Supports: CW, AMTOR ARQ and FEC 45-200 baud
RTTY (Baudot, ASCII 7 & ASCII 8) 45-200 baud
PACTOR listen mode for registered users.

INTCOM32.ZIP — INTCOM v3.2 - Shareware

Source: 8, 19

BayPac Mode: 3

Comments: INTCOM is a simply multimode program that performs well and is easy to use. Except for an onscreen tuning scope (which is very useful), you won't find any other "Bells and Whistles".

Min Requirement: IBM compatible 286, Hercules or VGA, COM1 or COM2

Supports: CW, RTTY Baudot 45, 50, 75 & 100 baud, ASCII 110 baud
Sitor FEC Mode B

JVFAX71A.EXE — JVFAX v7.1a - Shareware

Source: 9, 19, 20

BayPac Mode: 3, 6

Comments: JVFAX is probably the most popular program for SSTV and WEFAX operation. It performs very well in both transmit and receive, and it is easy to use! When configuring JVFAX, select "HamComm" for the demodulator, and "Serial Audio" for the modulator. Also, make sure you have the correct base address and IRQ for your serial port!

Min Requirements: IBM compatible XT, VGA, DOS 3.0

Supports: SSTV Martin 1 and 2
Scottie 1, 2, and DX
Robot 72c, BW 8s, 16s, and 32s
WR24/128, WR48/128, WR48/256, and WR96/256
WR120 and WR180
FAX Wefax 288, Wefax 576, Msat CH1, and Msat CH2
HamColor, Ham 288b, Metr SN, and Metr NS
NOAA NS, NOAA SN, Color 240, and H288/120

MSCAN202.ZIP | Micro Scan v2.02 - Shareware

Source: 10, 19, 26 | BayPac Mode: 3, 6

Comments: Micro Scan is a feature packed SSTV and FAX program. Unlike most programs, Micro Scan allows you to load or edit a picture at the same time that you are receiving or transmitting! Mico Scan supports a mouse, and it comes with a built in paint program, and spectrum display to aid in tuning. A connection to the PC speaker is required for transmitting, and the unregistered shareware version only supports those modes marked with an asterisk.

Min Requirements: IBM compatible 386DX, SVGA, COM1 or COM2 only

Supports:

SSTV Martin M1* and M2
Scottie S1*, S2, and DX
BW SSTV 7.2*, 8*, 16*, and 32
Wraase 24*, 48*, and 96
FAX BW 60, 90, 120*, 180, and 240 LPM
Colour 120, 180, 240*, and 360 LPM

MUBAY102.ZIP | Multi-user BayCom v1.02 - Shareware

Source: 11, 19, 22 | BayPac Mode: 1, 2

Comments: This is a Multi-user/multitasking packet radio program that allows multiple operators to access different areas of the system at the same time! Installation of this program is a little more complex than terminal type programs, but it is fairly easy to get it up and running. The Personal Mail System (PMS) is disabled in the shareware version.

Min Requirements: IBM compatible XT, 640k, Mono, DOS 3.0
Uses the TFPCX ax.25 driver (see comments under TFPCX210.EXE).

Supports: 300 and 1200 baud packet. UUencode/UUdecode of files. YAPP, 7Plus and ASCII file transfer. A script language, file/notepad editor and directory browser are also included.

PC HF FAX | PC HF FACSIMILE v7.0 - Software Systems Consulting

Source: 14 | BayPac Mode: 5, 6

Comments: Professionally written, documented, and supported software. Performs very well in all modes. Transmit functions in almost all modes, including WEFAX!

Min Requirements: IBM compatible XT, 640K, CGA, DOS 2.1

Supports:

CW, RTTY (Baudot, ASCII 7 & ASCII 8) 45-100 baud
ARQ, FEC, NavTex 75 and 100 baud
Raw Hex 45-100 baud
Fax 480 HAM
WEFAX 20-240 LPM
PRESS FAX 20-240 LPM and 480 HAM

PC SSTV

Source: 14

PC SSTV v5.2 - Software Systems Consulting

BayPac Mode: 5, 6

Comments:

Professionally written and well supported software. Supports most SSTV modes and offers the user the ability to define new (custom) modes.

Min Requirements:

IBM compatible 286, 640K, VGA, DOS 2.1

Supports:

SSTV 2, 24, 36, and 72 sec Color
8, 12, 24, and 36 sec BW
Scottie 1, and 2
Martin 1, and 2
AVT 90, 94, and 125

PKTMON12.ZIP

Packet Monitor v1.2 - Shareware

Source: 12, 19

BayPac Mode: 3, 6

Comments:

This is a receive only program used for monitoring HF and VHF packet radio traffic. PKTMON logs received data into separate files based on source and destination add An on-screen tuning indicator aids in tuning.

Min Requirements:

IBM compatible 286

Supports:

300 and 1200 baud packet

PROSKAN.ZIP

Source: 13, 19, 20
21

Pro Scan v3.01 - Shareware

BayPac Mode: 3, 6

Comments:

PROSKAN allows you to receive and transmit SSTV images. FAX modes are also provided, but they are for reception only. PROSKAN is easy to configure, it supports a mouse, and it has a built-in paint program for editing or creating images. The unregistered shareware version of PROSKAN disables or limits a lot of the programs better features.

Min Requirements:

IBM compatible 386DX, 640k, SVGA, DOS 4.0

Supports:

SSTV Martin 1 & 2, Scottie 1, 2, DX, & DX2
BW 24, & 36, AVT 24, 90, & 94, J-120
FAX WEFAX (576 IOC) 40-240 LPM

SP650A.EXE

Source: 15, 19

Super Packet v6.5a - Shareware

BayPac Mode: 1, 2

Comments:

This is the last shareware release of Super Packet. Installation and setup of this program is complex, however it is easy to use once you get it up and running.

Min Requirements:

IBM compatible XT or Atari ST, 640K, Mono
Uses the TFPCX ax.25 driver (see comments under TFPCX210.EXE

Supports: 300 and 1200 baud packet. BBS capability with 7PLUS and ASCII file transfer. Remote operations, scripts, and keyboard macros.

TFX28.LZH — TFX v2.8 - Shareware

Source: 19

BayPac Mode: 1, 2

Comments: TFX is a new RAM resident (TSR) program that receives and decodes packet transmissions. This program is similar to TFPCX in that it is used as the front end to other programs (like Graphics Packet, Super Packet, etc.). The TFX28.LZH archive file contains several different driver programs. The file TFX_PORT.COM supports BayPac 1200 baud modems, while the file TFX_PAR.COM supports the Tigertronics BP-96 (9600 baud) modem.

Min Requirements: IBM Compatible PC

Supports: 300, 1200, and 9600 Baud packet

TFPCX210.EXE — TFPCX v2.1 - Shareware

Source: 16, 19

BayPac Mode: 1, 2

Comments: TFPCX is a RAM resident (TSR) program that receives and decodes packet transmissions. TFPCX is almost always used as the front end to other programs (like Graphics Packet, Super Packet, etc.), however it can be set up to act as a VERY SIMPLE stand alone terminal program. Detailed instructions for setting up TFPCX can be found in the documentation files that accompany the program. Programs employing this driver work very well, but perhaps not quite as well as BayCom.

Min Requirements: 8MHz IBM compatible XT, 640k

Supports: 300 and 1200 baud packet. Graphics Packet, Multi-user BayCom, Super Packet, TOP, and TSTHOST are just a few programs that use TFPCX.

TOP151.ZIP — The Other Packet v1.51 - Shareware

Source: 17, 19, 22

BayPac Mode: 1, 2

Comments: The Other Packet (TOP) terminal program. TOP supports up to 10 operational channels, and provides keyboard macros and a screen saver.

Min Requirements: IBM compatible XT, 640K
Uses the TFPCX ax.25 driver (see comments under TFPCX210.EXE

Supports: 300 and 1200 baud packet. 7Plus, binary and ASCII file transfer protocol.

TSTHOST.EXE — TSTHOST v1.42 - Shareware

Source: 18, 19

BayPac Mode: 1, 2

Comments:	TSTHOST is a packet BBS/Server program that is loaded with features! Up to eight communications channels are supported, and all channels may be active with file transfer or Personal Mail System (PMS) activity at the same time!
Requires:	IBM compatible XT with 640K, EGA or VGA Uses the TFPCX ax.25 driver (see comments under TFPCX210.EXE
Supports:	300 and 1200 baud packet. 7PLUS, YAPP and ASCII file transfer. Macro commands.

WINTNC11.EXE	Win TNC v1.01 - Shareware
Source: 11, 19, 22	BayPac Mode: 1, 2
Comments:	Packet radio for Windows 3.1, Windows For Work Groups, and Windows 95! WINTNC features simple installation, on-line configuration of install parameters, and on-line help. When configuring WINTNC, under "Port Configuration", set the parameter "Carrier Sense" to "2". Also, the author of WINTNC reports that performance may be improved by setting the WALKSTEPDIV parameter to 64. Interim upgrades to WINTNC are being released as improvements are made. Check the WINTNC sources mentioned for these upgrades. The file WTNC101F.EXE is the current upgrade to WINTNC.
Supports:	300 and 1200 baud packet. Personal Mail System.

NOTES: 1) BayPac Mode #6 is receive ONLY, but works with almost all multimode programs.

2) Only the shareware version of BayCom (v1.4) is available on the ARRL BBS.

SOURCES
=======

1) Advanced Elect Applications
P.O. Box C2160 — Voice (206) 774-5554
2006 196th Street SW — Fax (206) 775-2340
Lynnwood, WA 98036 — Compuserve 76702,1013

2) BayCom v1.6
Tigertronics, Inc.
400 Daily Lane
P.O. Box 5210 — Voice (541) 474-6700
Grants Pass, OR 97527 — Fax (541) 474-6703

3) EZSSTV.ZIP
John Langner WB2OSZ
115 Stedman St. # E
Chelmsford, MA 01824-1823 — Voice (508) 256-6907

4) FBB515.ZIP
Jean Paul Roubelat
6, rue George SAND
31120 ROQUETTES
FRANCE

5) GP161B.ZIP

Ulf Saran
DH1DAE
Veit-StoB-Str. 36
57076 Siegen

6) GSHPC12.ZIP
G.Szabados-Hann DL4SAW
Am Zundhutle 7a — Voice 0721 / 47-53-19
76228 Karlsruhe / Germany — Fax 0721 / 47-53-19

7) HAMCOM31.EXE
W. F. Schroeder
Augsburger Weg 63
D-33102 Paderborn
Germany — Email schroeder.pad@sni.de

8) INTCOM32.ZIP
P.M. Haringsma (PA3BYZ)
dr. Hattinkstr. 13
8563 AC Wijckel
The Netherlands

9) JVFAX71A.ZIP
Eberhard Backeshoff
DK8JV
Obschwarzbach 40a
D-40822 Mettmann
Germany

10) MSCAN202.ZIP
CombiTech
Morelstraat 60
3235 EL Rockanje — Voice +311814-4252
The Netherlands — Fax +311814-4252

or North American agent:

HAMVision NA
P.O. Box 5073 — Voice (704) 636-3308
Salisbury, NC 28147-7806 — Fax (704) 636-3308

11) MUBAY102.ZIP
WINTNC11.ZIP
Jon Welch
50 Quarrydale Road
Sutton In Ashfield,
Notts
NG17 4DR
Great Britain — Email jon@g7jjf.demon.co.uk

12) PKTMON1.ZIP
Pawel Jalocha
Rynek Kleparski 14/14a
PL-31150 Krakow

13) PROSKAN.ZIP
Maynard A. Philbrook jr.
520 Pleasant St.
Willimantic, CT 06226 — Voice (203) 456-1167

14) Software Systems Consulting — Voice (714) 498-5784
615 S. El Camino Real — Fax (714) 498-0568
San Clemente, CA 92672 — BBS (619) 259-5554

15) SP650A.EXE
Orfeo DiBiase, NS8M
640 E. 8th St.
Salem, OH 44460 — Email ba350@yfn.ysu.edu

16) TFPCX210.EXE
Rene Stange
O.-Grotewohl-Ring 34
15344 Strausberg

17) TOP151.ZIP
F5NZE
Bernard MEBS
20 Rue Des Roses
67120 DUPPIGHEIM
TEL/FAX 88507120

18) TSTHOST.ZIP
IK1GKJ
Mario Travaglino
Via Trieste, 10
28069 Trecate (No) - ITALY

19) ARRL BBS (860) 594-0306

Internet links

20. ftp://ftp.funet.fi/pub/ham/fax_sstv/
21. http://www.mindport.net/~jamie/
22. http://members.aol.com/g7jjf/index.html
23. http://www.vectorbd.com/bfd/f6fbb/
24. http://www.vectorbd.com/bfd/packterm/
25. http://www.ultranet.com/~sstv/
26. http://www.pi.net/~ctech/

NOTE: Ftp links are available for most of these programs on the download page of our web site. (http://www.tigertronics.com)

SHAREWARE - DOING YOUR PART!

Most of the programs on this list are Shareware. Simply put, Shareware means that you are welcome to download and use the program at no charge for a limited trial period. After the evaluation period, if the program meets your needs and expectations and you plan to continue using it, you are asked to send a small registration fee to the author.

Some of these programs represent "man years" of effort and rival their commercial counterparts. By marketing these programs as shareware, the authors are able to bring you outstanding products at rock bottom prices. But, Shareware only works if you honor your part of the agreement. Tigertronics encourages you to register the software you choose to keep and support the authors in their work. Register today!

INCOMPATIBLE PROGRAMS

AMTOR.EXE — Older version of TOR306.ZIP (See TOR306.ZIP)

ATFAX.ZIP — Requires special interface.

EMBAYCOM.ZIP — Requires modified "HamComm" interface. This program allows a "HamComm" interface to run BayCom software. This program is of little use since the BP-2M runs BayCom directly.

L2PCX.EXE — Supports serial FSK modem only.

PAC020.ZIP	Requires special interface.
PCTOR.ZIP	Older version of TOR306.ZIP (See TOR306.ZIP)
PD-203.ZIP	Requires special interface.
PROCW.ZIP	Requires external tone decoder on receive.
RTTY12G.EXE	Requires special interface.
SSTVFAX2.ZIP	Requires special interface.
TOR306.ZIP	Requires special interface.
TTY37.ZIP	Requires special interface.
VESTER_M.ZIP	Complex setup/operation - We couldn't get it to run.
WXMAN2.ZIP	Sound Card only.
W95SSTV.ZIP	Sound Card only.
SSTVBL.ZIP	Sound Card only

Chapter 2
Theory/Design

"9600-Ready" Radios: Ready or Not?

As VHF/UHF FM transceiver manufacturers take the plunge and offer new radios that claim to be ready for 9600-baud packet operation with little or no modification, users are wondering if they really work. The ARRL Lab takes an objective look.

By Jon Bloom, KE3Z
ARRL Senior Engineer

The increasing popularity of 9600-baud packet, sparked by the existence of several 9600-baud satellites and by the frustration many packet users feel with 1200-baud terrestrial packet, has induced several equipment manufacturers to include 9600-baud capability in their new VHF and UHF transceivers. As with all aspects of radio performance, it's likely that not all radios perform equally well at 9600 baud. To discover what differences there are in the 9600-baud performance of these radios, the ARRL Lab has developed techniques to test radio performance with 9600-baud signals. In this article, we present a brief description of those techniques, along with test results for radios that have recently appeared in Product Review.[1,2] A companion article in a recent issue of *QEX* provides the full technical details of the test setup and software.[3] Future reviews of 9600-baud radios will include measurements made using these techniques.

Background

Packet operation at 9600 baud imposes different requirements on the radio than does 1200-baud operation because the signals applied to the radio are very different. At 1200 baud, the binary data to be sent is used to generate an audio frequency-shift-keyed (AFSK) signal in the terminal node controller (TNC). This shifting audio tone, centered at 1700 Hz, is the signal applied to the transmitter's audio input. From the receiver, the audio signal is applied to the TNC, which demodulates the AFSK signal back into binary data. The benefit of this scheme is that the signal fits neatly into the frequency range of normal speech, so it can be sent through voice-grade radios.

9600-baud packet uses a different approach because it isn't feasible to generate an AFSK signal at 9600 baud that will fit within the voice frequency range. Instead, the binary data signal is encoded and filtered, then applied directly to the FM transmitter, producing a true frequency-shift-keyed (FSK) signal. The demodulated signal from the receiver is detected by simply determining whether the signal voltage is above or below zero—that is, whether the current level represents a 1 bit or a 0 bit. The resulting data stream is then decoded to recover the binary data originally sent.

The first 9600-baud packet system was developed by Steve Goode, K9NG.[4] In his scheme, the binary data from the TNC was first passed through a "scrambler" circuit. Despite its name, scrambling has nothing to do with encryption or hiding the data. It's simply a means of encoding the data to ensure that the number of 1 bits is essentially equal to the number of 0 bits in the transmitted data stream. Doing this eliminates any dc offset in the signal. For example, if we sent an unscrambled signal consisting of a number of bits at +1 V followed by a single bit at –1 V, with this

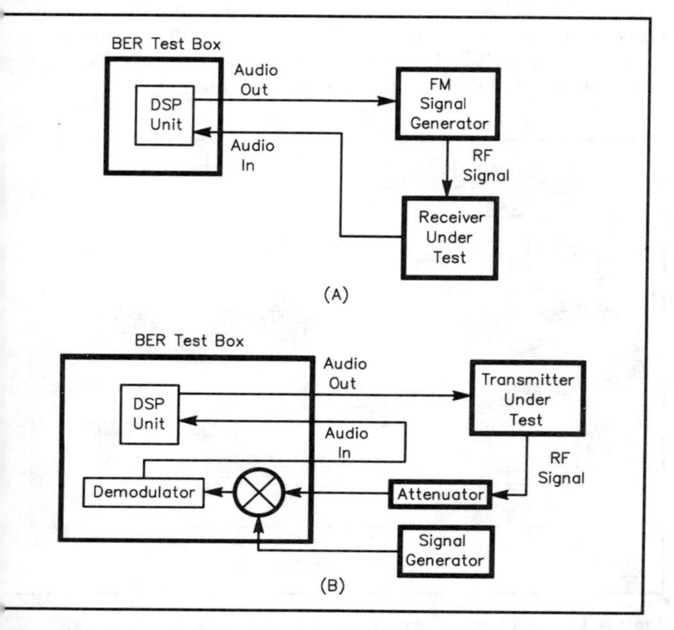

Figure 1—Bit error rate (BER) test setups used to test receivers (A) and transmitters (B).

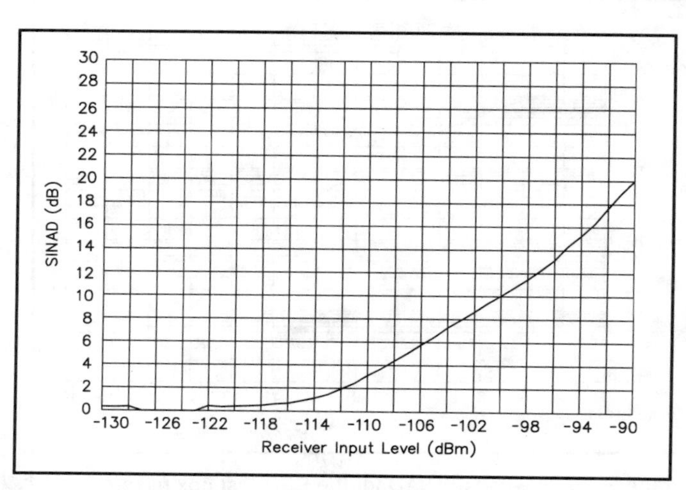

Figure 2—Measured SINAD versus input level of the BER test box mixer/demodulator.

sequence repeated a number of times, the signal would have a substantial dc component. Different data patterns would cause different dc components to be present. Scrambling eliminates this dc component, allowing us to use ac (capacitive) coupling at various points in our system since we don't need to preserve a dc part of the transmitted signal. It also means the system can tolerate a small frequency difference between the transmitter and receiver—a frequency difference that shows up as a constant dc offset.

K9NG's modem simply filtered the scrambled signal to eliminate harmonics of the square-edged data signal, then applied the filtered signal to the transmitter. But filtering the signal also causes each transmitted bit to spread out in time, overlapping the nearby bits. As long as the signal is transmitted, propagated and received without significant distortion (amplitude or phase error), this spreading of the bits isn't a problem. But if distortion is present, different parts of the transmitted bits get distorted differently, and the received signal no longer has the same shape as when it was transmitted. This can cause problems for the receiving 9600-baud modem.

To address this problem, James Miller, G3RUH, developed a 9600-baud modem design that, while compatible with the K9NG modem, generates a signal that is less susceptible to distortion.[5] In the G3RUH modem, the scrambled data stream is used to generate bit pulses that have a raised-cosine spectrum. Such a pulse is at its maximum amplitude at the center of the bit and, while the bit overlaps preceding and following bits, its amplitude is always zero at the centers of those bits.[6] That means that even if the signal is somewhat distorted, the part of a bit that overlaps nearby bits will still probably be near zero amplitude at the centers of those bits. Unless the distortion is severe, each transmitted bit should still be largely unaffected by the overlapped parts of the other bits.

G3RUH also recognized that it is possible to compensate for amplitude and phase distortion if the character of that distortion is known. For example, if you know the signal will pass through a receiver that attenuates high frequencies, you can boost the amplitude of the high frequencies in the transmitted signal. His means of providing compensation was to generate the transmitted pulses using a ROM look-up table, with settable jumpers to select the particular pulse shape from the ROM. In a point-to-point link, where a single transmitter and receiver are used, this is an effective technique, allowing correction of the combined distortion of the transmitter and receiver. But in a system where a signal may be received by one of a number of receivers, only the transmitter distortion can be corrected this way.

The G3RUH system is the de facto standard 9600-baud packet modem, so we tested these radios for compatibility with G3RUH-type signals. These signals have a spectrum that is constant in amplitude from nearly dc to 2400 Hz, then gradually decreases, reaching –6 dB at 4800 Hz and diminishing to zero at 7200 Hz.

Some manufacturers are now selling 9600-baud TNCs that use single-chip 9600-baud modems. These operate using Gaussian minimum-shift keying (GMSK) waveforms, which are similar—but not identical—to the signals used by G3RUH modems. The two signal types are enough alike that G3RUH modems and GMSK modems are interoperable, although with a small performance penalty compared to using identical modems at both ends of the link. In general, the results you'll get from a given 9600-baud radio will be similar with either the G3RUH or GMSK modems, but specific radios may affect one type of signal more than another.

Testing Radio Performance

The best way to characterize the performance of a digital communication link is to get right to the bottom line: How accurately do the bits get through? The measure of this is the *bit error rate* (BER). The BER states how many of the bits sent through a system will be received incorrectly, on average. For example, if for each 1000 bits transmitted through the system, one bit is received incorrectly, the BER is 1/1000. For convenience, because the BER numbers are small (we hope!), they are usually written using scientific notation. In this example, the BER is 1×10^{-3}.

We can relate a given BER to real-world performance. In AX.25 packet, a single bad bit in a packet will cause the packet to be unusable. That means that if your average packet is, say, 1000 bits in length (a reasonable number), and the BER is 1×10^{-3} (one error in each 1000 bits), the average number of bad bits in each packet will be 1, and all of the packets will be unusable. Of course, the BER only gives an *average* rate of errors. In reality, some of the packets will have two or more bad bits; some will have none. So some of the packets will arrive intact...but not many! On the other hand, if the BER is 1×10^{-5} (one error in 100,000 bits) there will be, on average, one bad bit in every 100 packets. You might not even notice that!

In a well-designed system, the major factor that causes bit errors is noise. As the signal gets weaker, the signal-to-noise ratio drops and more of the transmitted bits are corrupted by noise, resulting in improper reception of the bits. Any distortion added by the transmitter and receiver causes even more bits to be received erroneously. It is this distortion added by the radio that we want to measure.

The test setups we used are shown in Figure 1. For receiver testing, the 9600-baud test signal is used to modulate a Marconi 2041 signal generator, which provides distortion-free modulation of the signal. In transmitter testing, transmitted signals from the radio being tested are mixed down to a low IF and demodulated using a wide-band, low-distortion demodulator built by the ARRL Lab and described in the companion *QEX* article.

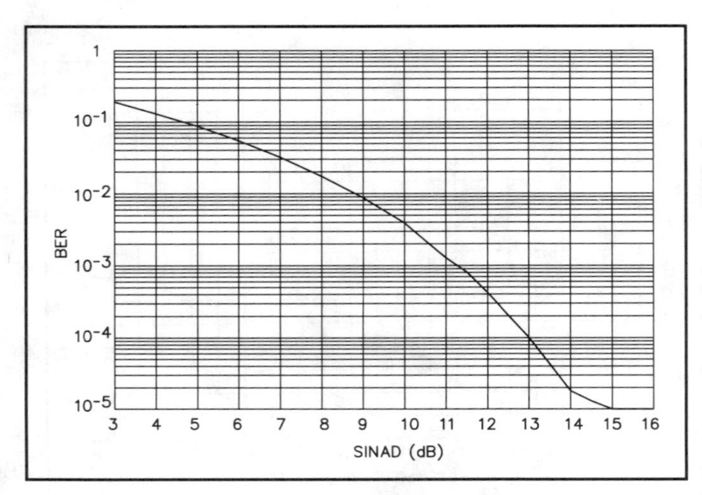

Figure 3—BER versus SINAD for the BER test box mixer/demodulator.

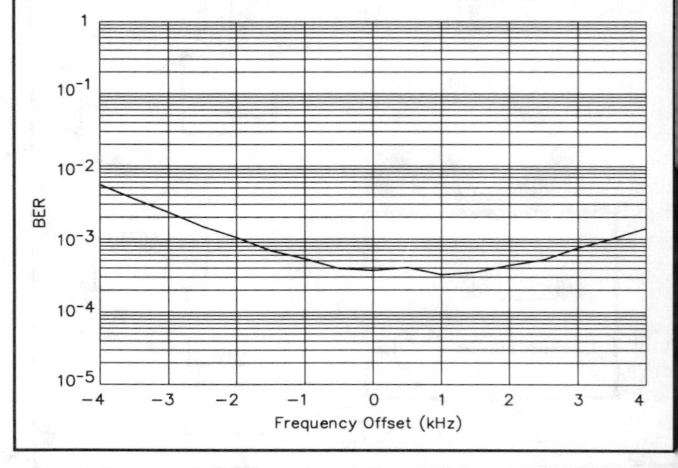

Figure 4—BER versus frequency offset for the BER test box mixer/demodulator.

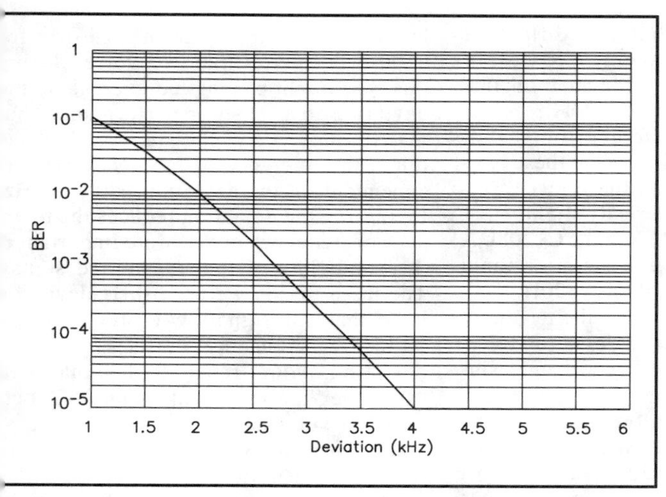

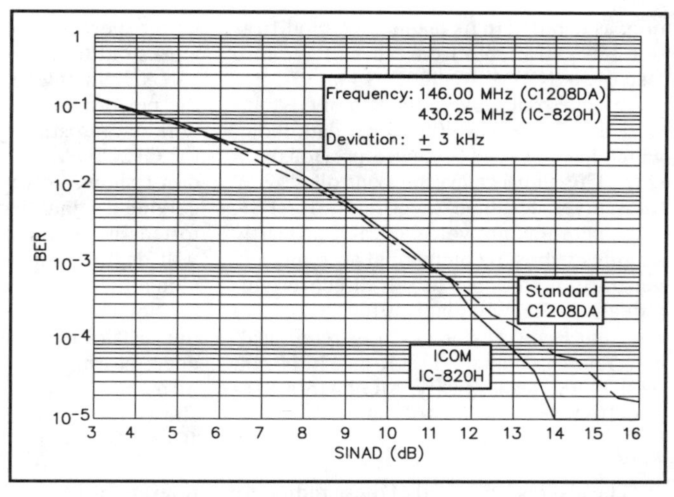

ᵎigure 5—BER versus deviation for the BER test box mixer/ demodulator.

Figure 6—BER versus SINAD for the ICOM IC-820H and Standard C1208DA receivers.

The BER test box generates a pseudo- andom stream of bits using the same algo- ithm implemented by the scrambler cir- uit in the K9NG and G3RUH modems. ʾhis data stream is formed into a G3RUH- ype signal which is applied to the signal enerator (for receiver testing) or the trans- nitter under test (for transmitter testing). After passing through the system and being demodulated, the signal returns to the BER est box. The test box inspects the incom- ng signal and compares it to the signal that vas sent. To accomplish this, the test engi- eer adjusts a delay value in the BER test ox that matches the delay through the sys- em. If there are no errors induced by the est system, the returned bit will always be he same (0 or 1) as the bit sent.

ounting Errors

Bit errors are by nature random. That neans that if you run the same number of its through the same system in two differ- nt tests, you will likely get different re- ults. But the average rate of errors should e consistent. That means that if you send nough bits through the system, you'll get epeatable results. So, how many bits do ou need to send? It turns out that you need o send enough bits to get a certain number f errors. How many errors you need to get epends on how confident you want to be f the results. If you get 100 bit errors, you an be 99% sure that the measured BER is vithin a factor of about 1.3 of the value you vould get if you sent an *infinite* number of its through the system (the "true" BER). If ou get only 10 errors, you can be 95% sure nat the BER is within a factor of 2 of the ue BER.[7] Our test technique then, is to end up to 1 million bits through the sys- em. That lets us measure BERs of 1×10^{-4} 100 errors) or more with a high degree of onfidence and BERs of 1×10^{-5} (10 errors) ith reasonable confidence. The BER test ox sends bits through the system 100,000 a time, and it always sends at least 300,000, but if it gets more than 100 errors before reaching 1 million bits, it stops there and calculates the BER using the number of bits sent.

Tests Performed

To find out what effect the radios have on the signal, we need to test them using signals that are somewhat noisy. That way, the errors added by radio distortion become quite apparent. We wanted to measure each receiver at specific signal-to-noise ratios. Since each unit exhibits a unique sensitivity, a particular signal-to-noise ratio will occur at a different input power level for each receiver. To standardize the signal-to-noise ratio, we first measured the SINAD of the receiver at its data output using a 1-kHz tone and 3-kHz deviation. We swept the input power level from –130 dBm to –90 dBm, resulting in a curve like that of Figure 2. (Figures 2 through 5 are measurements of the BER test box's mixer/demodulator, which acts like a receiver. This circuitry is used to test transmitters, and testing the BER test box as though it is a receiver shows us the performance of the transmitter test setup.) From that curve for any particular radio we can establish the input signal level needed to produce a particular SINAD.

For receivers, we performed three swept BER tests and one single-state test. First, we measured the BER using a 3-kHz peak deviation and a signal at the nominal operating frequency of the receiver. The input power level was stepped to measure BER at SINADs from 3 to 16 dB. This produces a curve like that of Figure 3. Next, we set the input level to that needed for a 12-dB SINAD and swept the signal frequency from 4 kHz below the nominal operating frequency to 4 kHz above, producing a curve like that of Figure 4. This tests the receiver's tolerance of frequency error. Our final swept measurement was made at the operating frequency but with the peak deviation stepped from 1 kHz to 6 kHz, to test the receiver's tolerance of over- and under-deviated signals (Figure 5). Our final test was to measure the BER on-frequency with 3-kHz deviation and a –50 dBm input signal level. This is a *strong* signal that should result in an extremely low BER.

Transmitter testing was somewhat simpler. Using the Lab-built mixer/demodulator in the BER test box, we attenuated the transmitter output to get a 12-dB SINAD from the test box mixer/demodulator and measured the BER using 1 million bits. We then increased the transmitter signal into the mixer by 30 dB and measured the BER again. The latter test should result in essentially no bit errors if the transmitter is distortion-free.

As you'll see, the results of our tests are not particularly encouraging. For this reason, we decided to measure the amplitude and phase responses of all of the radios, to see if we could find out why some radios produced such poor BER results. The system used to make the response measurements has been described previously in *QEX*.[8]

To provide a final check of our testing, we spot checked some of our measurements with a G3RUH modem, using its BER test mode to send and receive the data signals. The results showed a small, consistent improvement in measured BER with the G3RUH modem compared to our test set. This is accounted for by the input low-pass filter of the G3RUH modem. It cuts off at a lower frequency than the filter in our test set, which removes some of the noise. The signal-to-noise ratio—and the BER—is therefore improved slightly.

Test Results

We tested four previously reviewed 2-meter FM radios: an ICOM IC-281H, Kenwood TM-251A, Standard C1208DA and a Yaesu FT-2500M. The FT-2500M requires modification for 9600-baud operation, as detailed in its manual. (A modification kit is available from Yaesu.) We also tested an ICOM IC-820H 144/430-MHz multimode transceiver. And finally, we included a TEKK KS-900 70-cm data radio, which has not been reviewed previously in *QST*. This unit is a crystal-controlled, 2-W transceiver designed specifically for FSK data operation, not voice. (TEKK primarily supplies radios to commercial accounts and generally perfers not to sell direct to individuals. TEKK will gladly refer prospective Amateur Radio customers to its retail dealers. You can contact TEKK at 226 Northwest Pkwy, Kansas City, MO 64150; 816-746-1098 or fax 816-746-1093.)

Receiver Test Results

The test results for all of these radios are shown in Table 1, in the format we'll use to report BER measurements in future reviews. For receivers, we've selected three signal levels at which to report the BER exhibited by each receiver. The 12-dB SINAD BER is used as a reference for other measurements and is also a level at which even an ideal receiver will exhibit a BER of greater than 1×10^{-4}. This column of Table 1 shows gross differences in weak-signal performance between receivers. The 16-dB SINAD level is one at which excellent receivers should show a BER of less than 1×10^{-5}. As Table 1 shows, some receivers do show such performance, some don't. Finally, a measurement of BER with a –50 dBm input signal level should result in a BER of less than 1×10^{-5} from any receiver that is even marginally usable, and all of the receivers tested met this criterion.

One should be careful about making too much of relatively small differences between receivers. For example, Figure 6 shows the full swept BER versus SINAD for two of the tested receivers: the ICOM IC-820H and Standard C1208DA models. These two units show the greatest difference in their receiver BER performance against noise, with the IC-820H having the better performance, but the difference works out to be equal to only a few dB of SINAD. So, it would be wrong to think that the performance of the IC-820H receiver will be dramatically different from that of the C1208DA in practice. What these tests don't show, however, is the effect of receiving a signal that is already slightly distorted. In general, the BER of the C1208DA will degrade more from a distorted input signal than will that of the IC-820H.

Not shown in Table 1 are the results of our tests of frequency and deviation tolerance. Normally, we will report only significant results. For example, Figure 7 shows the BER versus frequency for the IC-820H and the C1208DA. Of all the radios tested, the IC-820H was the most tolerant of frequency error. This tolerance can be helpful when you are trying to work the 9600-baud packet satellites, with their constantly changing, Doppler-shifted downlink frequency. The C1208DA, on the other hand, can accept only a relatively small frequency error before performance degrades substantially. This, too, has its positive side, as it indicates that the bandwidth of the C1208DA receiver is relatively narrow, leading to improved adjacent-channel rejection.

All of these radios not only tolerate substantial overdeviation, they perform *better* at higher deviations. Figure 8 shows the BER versus deviation of the Yaesu FT-2500M (after modification), which is representative in that the BER decreases (improves) as deviation is increased. All of the radios showed substantially improved performance at 5-kHz deviation compared to the usual 3-kHz deviation. This really shouldn't be a surprise. Consider that in FM voice communication, we use a deviation of 5 kHz to carry voice signals that have a frequency content of up to about 3 kHz. In 9600-baud packet, our spectrum goes up to 7.2 kHz (albeit at fairly low levels toward the top of the spectrum) and the deviation is only 3 kHz. Our modulation index is therefore much lower for 9600-baud packet than for voice. Seeing this, you might be tempted to crank up the deviation to get better performance, but that's not generally a good idea. For one thing, the tolerance of frequency error is dependent on the deviation being 3 kHz. Increasing the deviation reduces the tolerance of frequency error. More important is the bandwidth of the transmitted signal. The main reason we use 3-kHz deviation is so that our signals will stay within a 20-kHz-wide channel. If you increase the deviation, your transmitted signal will "slop over" into adjacent channels. A better fix if your system isn't performing as you'd like is to install a better antenna or to increase transmitter power. However, for point-to-point network backbone links running in wide channels (such as the 100-kHz-wide channels at 70 cm), where the frequency offset can be controlled since there are a limited number of stations using the link, it may well make sense to increase the deviation to 4 or 4.5 kHz.

The bottom line for these receivers is that they all are capable of receiving 9600-baud packet signals. There are performance differences, however, and these differences may become more significant when distorted, real-world signals are being received.

One thing about the IC-820H should be noted. The impedance of its data output is very high—nearly 50 kΩ. This means that the modem or TNC connected to the IC-820H must have a high-impedance input. It especially must have an impedance that is constant across the frequency range of the 9600-baud audio signal. The G3RUH modem does *not* have a constant impedance across this range. At low frequencies, its input impedance is near 100 kΩ, but its impedance decreases at higher frequencies. For example, its impedance at 4 kHz is about 55000–*j*45000 Ω. With this difference in impedance across the frequency range, the effect of the load is to distort the

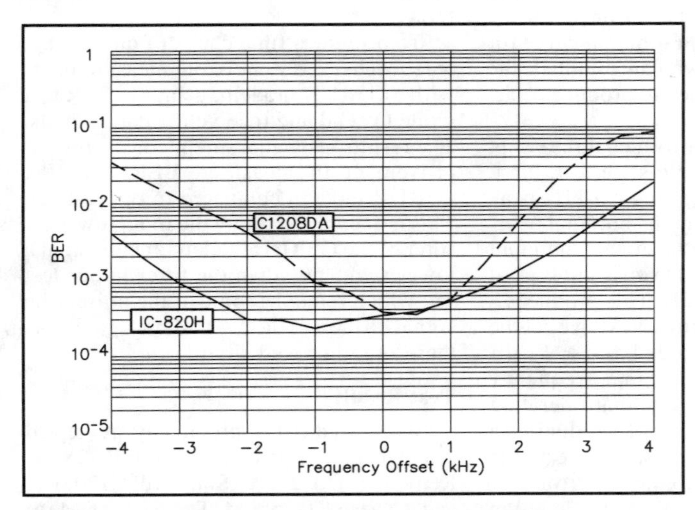

Figure 7—Frequency tolerance curves for the ICOM IC-820H and the Standard C1208DA receivers.

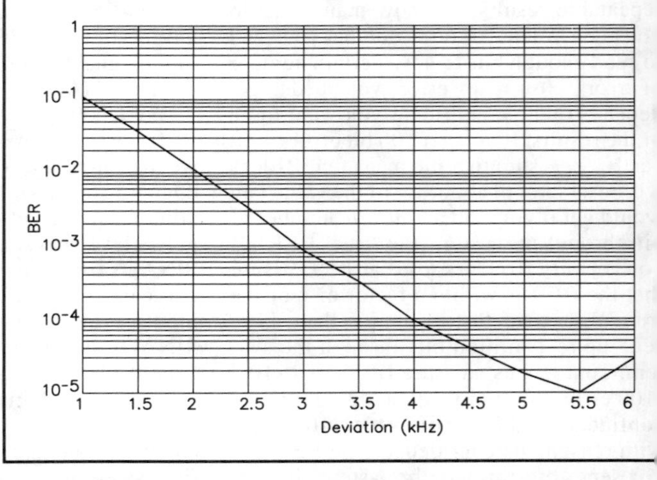

Figure 8—BER versus deviation for the Yaesu FT-2500M (modified) receiver.

Table 1
Measured Performance of 9600-Baud Radios

Model	Receiver: 12-dB SINAD level (dBm)	Receiver: BER at 12-dB SINAD	Receiver: BER at 16-dB SINAD	Receiver: BER at –50 dBm	Transmitter: BER at 12-dB SINAD	Transmitter: BER at 12-dB SINAD +30 dB
2-meter FM radio:						
ICOM IC-281H	–113.2	4.8×10^{-4}	$<1\times10^{-5}$	$<1\times10^{-5}$	7.4×10^{-3}	1.1×10^{-3}
Kenwood TM-251A	–114.3	3.6×10^{-4}	1.3×10^{-5}	$<1\times10^{-5}$	2.2×10^{-3}	2.9×10^{-4}
Standard C1208DA	–108.8	3.9×10^{-4}	1.7×10^{-5}	$<1\times10^{-5}$	3.7×10^{-3}	7.0×10^{-4}
Yaesu FT-2500M	–113.3	8.6×10^{-4}	$<1\times10^{-5}$	$<1\times10^{-5}$	4.9×10^{-3}	3.8×10^{-4}
144/430-MHz multimode radio:						
ICOM IC-820H	–113.0	2.7×10^{-4}	$<1\times10^{-5}$	$<1\times10^{-5}$	5.1×10^{-4}	$<1\times10^{-5}$
440-MHz data-only radio:						
TEKK KS-900	–110.4	2.6×10^{-4}	$<1\times10^{-5}$	$<1\times10^{-5}$	3.8×10^{-4}	$<1\times10^{-5}$

signal coming from the IC-820H. This can be seen in Figure 9, which shows the BER versus noise of the IC-820H both with and without the G3RUH modem loading the receiver output. The effect isn't large in this case, but it is present. This suggests that you should check the output impedance of the radio and the input impedance of the modem or TNC before connecting them together. If the radio's output impedance is higher than the TNC's input impedance, you may want to add a simple op-amp or one-transistor buffer between the radio and the modem.

Transmitter Test Results

Although there are some differences in receiver performance, it is in transmitter performance that we see huge differences among the radios tested. Our measurements were made with both noisy (12-dB SINAD) and strong (12-dB SINAD plus 30 dB) signals. As a reference, we measured the BER at these levels using a signal generator as the "transmitter." The near-perfect signal from the generator produced a BER of 3.33×10^{-4} at 12-dB SINAD, and no bit errors (BER < 1×10^{-5}) at a level 30 dB greater. Each of the transmitters tested gave higher BERs than the reference signal generator, as you would expect. The numbers are shown in Table 1. Remarkable among these were the IC-820H multimode radio and the KS-900 data radio. Their transmitter performance was nearly as good as that of the reference signal generator. None of the 2-meter FM transmitters achieved a low BER, even with a strong signal! This is quite troubling, if not entirely unexpected.

The main difficulty is in the low-frequency response of the transmitters. Unlike voice signals, 9600-baud packet signals contain significant components at frequencies below 100 Hz. This poses a problem for a synthesized transmitter, as explained in the sidebar, "Making Transmitters Speak 9600 Baud."

While we can understand the difficul-

The Bottom Line: How Good is Good Enough?

The values in Table 1 tell the story of how each of the tested radios performs. But what, ultimately, do these numbers mean in terms of on-the-air performance? That's hard to say, specifically. What we *can* say is that lower numbers mean better performance. (Be sure to take the exponent of 10 into account when comparing values.) The strong-signal numbers (at –50 dBm for receivers and at 12-dB SINAD plus 30 dB for transmitters) are perhaps the best indicators of the general suitability of the radio for 9600-baud operation. In these columns, a radio should be able to exhibit a BER of 1×10^{-5} or less. As Table 1 shows, all of the receivers we tested do so. But only two of the transmitters produce this low BER. All of the other transmitters produce a BER that is at least 10 times higher. In side-by-side comparisons over actual radio links, these high-BER transmitters are likely to perform substantially less well than the transmitters that achieved a BER of less than 1×10^{-5}.

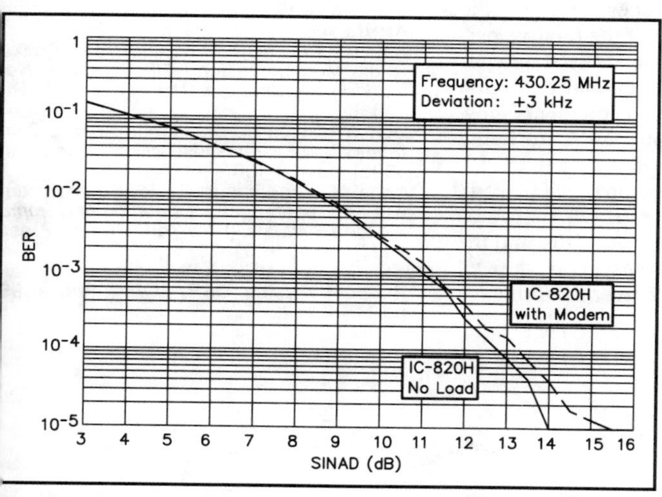

Figure 9—BER versus SINAD for the IC-820H, both without a load on the receiver data output and with a load consisting of a G3RUH modem.

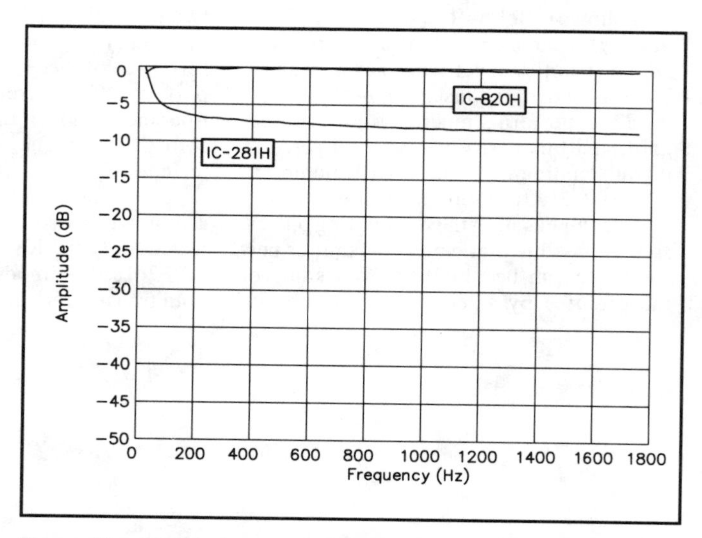

Figure 10—Amplitude response (to 1800 Hz) of the IC-281H and IC-820H transmitters.

Making Transmitters Speak 9600 Baud

Properly modulating an FM transmitter with a G3RUH-type signal requires a transmitter modulation response that is flat in amplitude and linear in phase from a very low frequency (near dc) up through about 7200 Hz. This is easy to do if the oscillator being modulated is a free-running VCO or VXO, but can be quite difficult if the oscillator is embedded in a phase-locked loop (PLL). And these days, the frequency-controlling oscillator in most transceivers is part of a PLL.

The problem arises because the PLL will try to "correct" any offset between the oscillator's operating frequency and the programmed frequency of the PLL. A PLL contains a low-pass loop filter, and only those frequency components of the modulating signal that fall within the loop filter will be affected by the PLL. Therefore, some of the modulating signal goes through untouched and some doesn't. The result, when the signal is received at the other end of the link, is a distorted signal that may be unusable by the receiving TNC.

Another approach to modulating the PLL is to modulate the reference oscillator. In this case, the PLL forces the VCO to track the change in the reference. This time, only those frequency components that fall *within* the loop filter find their way onto the transmitted signal. That's no better. It might seem that a simple change to the PLL's low-pass loop filter could solve this problem, but that will lead to problems with loop lock-up time, capture range or noise performance. In some cases, loop instability might even result.

The obvious solution to this problem is to apply the low frequencies to the reference oscillator and the high frequencies directly to the VCO. This *two-point modulation* scheme can work, but you have to be very careful how you split the modulating signal into its high- and low-frequency components so that the signal isn't distorted in the splitting. It isn't easy.

The ideal solution would be to modulate a VXO, then mix, or heterodyne, that signal with the output of an unmodulated synthesizer to produce the signal to be transmitted. This requires additional circuitry, of course. But consider an all-mode (CW/SSB/FM) radio like the IC-820H. To generate the SSB signal, such a radio already generates a fixed-frequency signal and then heterodynes it up to the output frequency. So the additional complexity needed for good 9600-baud modulation is already largely present. Of course, you may not want all-mode capability. If you are interested only in packet operation, the cost of the additional circuitry that an all-mode radio uses to do CW and SSB may be prohibitive.

The radios we tested for this article aren't the only ones on the market, of course. Elsewhere in this issue (Product Review) we report on an Azden PCS-9600D FM/packet transceiver, and other manufacturers are bringing out new data-radio designs quickly.

ties of the manufacturers who are trying to add 9600-baud capability to their voice radios, our understanding doesn't change the fact that the transmitter performance of most of these radios makes it difficult to successfully use 9600-baud packet. This is particularly true when the station at the other end is using a receiver that exhibits marginal performance.

Figure 10 shows the amplitude response of two of the tested transmitters, the IC-820H and the IC-281H. The lowest frequency point plotted on the graph is 22 Hz. The IC-820H is essentially flat from 22 Hz up, while the IC-281H significantly boosts the lower frequencies. Of equal concern is the variation in group delay with frequency (not shown). Below 100 Hz, the IC-281H displays a large variation in group delay, which is not unusual when large changes in amplitude response are present. The IC-820H performs much better in this regard, and this performance is mirrored in its outstanding transmitter BER numbers.

The G3RUH modem provides a way of precompensating the modulating pulses, so it is possible that better transmitter performance can be obtained from some of these radios by selecting an appropriate pulse shape from the ROM in the modem. This requires some experimentation. The G3RUH modem generates a test signal you can use for this purpose in its BER test mode. By receiving the transmitter's signal with a known good receiver, you can select from among the available pulse shapes to get the one that delivers the best performance. But some newer 9600-baud TNCs don't have this facility. TNCs built using single-chip 9600-baud modems *depend* on proper performance of the transmitter.

Conclusion

In future reviews of 9600-ready radios, we will present data in the form of Table 1. From the three receiver BER values and two transmitter BER values for each radio you can get a good idea of the relative performance of these products. We'll only report in *QST* on frequency and deviation tolerance results if the measurements show something unique about the radio being tested.

The readiness of the crop of 9600-ready radios we tested is mixed. Of the radios tested, only the KS-900 data radio and the IC-820H multimode transceiver showed both transmitter and receiver performance that will result in efficient packet operation at 9600 baud. Of the 2-meter FM transceivers, receiver performance was acceptable in all cases—if not always outstanding—but transmitter performance was disappointing.

Notes

[1]Steve Ford, WB8IMY, "*QST* Compares 2-Meter FM Mobile Transceivers," *QST*, Jan 1995, pp 70-76.

[2]Steve Ford, WB8IMY, "ICOM IC-820H VHF UHF Multimode Transceiver," *QST*, Ma 1995, pp 80-83.

[3]Jon Bloom, KE3Z, "Measuring 9600-Baud Radio BER Performance," *QEX*, Mar 1995, pp 16-23.

[4]Steve Goode, K9NG, "Modifying the Hamtronics FM-5 for 9600 bps Packet Operation," *Fourth ARRL Amateur Radio Computer Networking Conference*, (Newington: ARRL, 1985).

[5]James Miller, G3RUH, "9600 Baud Packet Radio Modem Design," *7th Computer Networking Conference*, (Newington: ARRL 1988).

[6]Leon W. Couch III, *Digital and Analog Communication Systems*, (New York: Macmillan 1993), pp 179-180.

[7]Michel C. Jeruchim, Philip Balaban and K. Sam Shanmugan, *Simulation of Communication Systems*, (New York: Plenum Press 1992), pp 498-501.

[8]Jon Bloom, KE3Z, "Measuring System Response using DSP," *QEX*, Feb 1995 pp 11-23.

By Tim Riley; Dennis Bodson, W4PWF; Stephen Rieman; and Teresa G. Sparkman

A Comparison of HF Digital Protocols

The world of HF digital communication is a confusing mix of protocols. Which offer the best performance?

In recent years our HF digital subbands have become Towers of Babel with several protocols competing for dominance. The familiar list includes AMTOR, G-TOR, CLOVER, PACTOR and PACTOR II.[1] Each mode has its enthusiastic supporters, but which ones are most efficient when it comes to transferring information?

The task of evaluating HF digital protocols is more than just an academic exercise. Our government has a stake in determining the answers because HF digital operators have traditionally supplied a large reservoir of backup communications services and operators during emergencies. In the interest of tapping this resource, the National Communications System (NCS) sponsored the National Telecommunication and Information Administration (NTIA) to test various HF modems and establish a technical baseline of standardized performance data on HF modem protocols. The Federal Emergency Management Agency (FEMA) will use those test results to determine which protocols could be used to serve as an interchange with the amateur community in the event of a national emergency. This article reflects the results of tests conducted at NTIA's Institute for Telecommunication Sciences (ITS) laboratories in Boulder, Colorado.

Solving the Propagation Problem

Many over-the-air tests have been conducted on various HF modem/protocol combinations. Atmospheric conditions vary so much, however, that it's impossible to draw solid conclusions. Over-the-air test results can change from day to day, or even hour to hour. To obtain scientifically valid information, you need stable conditions. Unfortunately, Mother Nature refuses to guarantee the stability of HF propagation!

So, we turn to the laboratory, where it's possible to create elaborate *simulations* of HF propagation paths. The engineers at ITS have developed an automated test bed for use in conducting controlled laboratory testing of modems and their protocols. Using this test bed, ITS subjected the protocols to a repeatable set of simulated propagation paths over a wide range of signal-to-noise ratios (S/Ns).[2]

Data transfers were performed at various S/Ns for each of the five protocols in their Automatic Repeat Request (ARQ) mode. All files received were checked for errors to determine the validity of the tests. The *throughput*, a measure of the data transfer rate, was measured for each protocol under simulated ionospheric conditions.

To conduct our tests we obtained HF modems manufactured by Kantronics, HAL Communications, Advanced Electronic Applications and SCS. However, specific models are not identified in this article. You'll see them referred to as modems A through D only. The intention is to evaluate protocols, not hardware. It is important to note that the use of specific hardware does *not* imply a recommendation or endorsement by the National Telecommunications and Information Administration, nor does it imply that the equipment used is necessarily the best available for this application.

Test Conditions

For each test, two identical modems (see Table 1), controlled and monitored by one 80486-based PC, were physically connected to each other by their audio-out/audio-in ports through two identical HF ionospheric channel simulators (Figure 1). Although the modems were both connected to the same PC, the operation of one was entirely independent of the other.

The modems were linked to the computer through the PC's serial communication ports (COM1 and COM2), except for the CLOVER modem, which plugged into the PC's ISA bus.

Table 1
HF Protocols Tested

Note: The AMTOR and PACTOR protocols were tested on two modems to determine if there were implementation differences in the results. The other three protocols were implemented by unique modems. Modems are identified simply as A, B, C and D.

Protocol	*Modem*	*Computer Interface*
AMTOR	A and B	Serial port
CLOVER	C	PC bus (plug-in card)
G-TOR	D	Serial port
PACTOR	A and B	Serial port
PACTOR II	A	Serial port

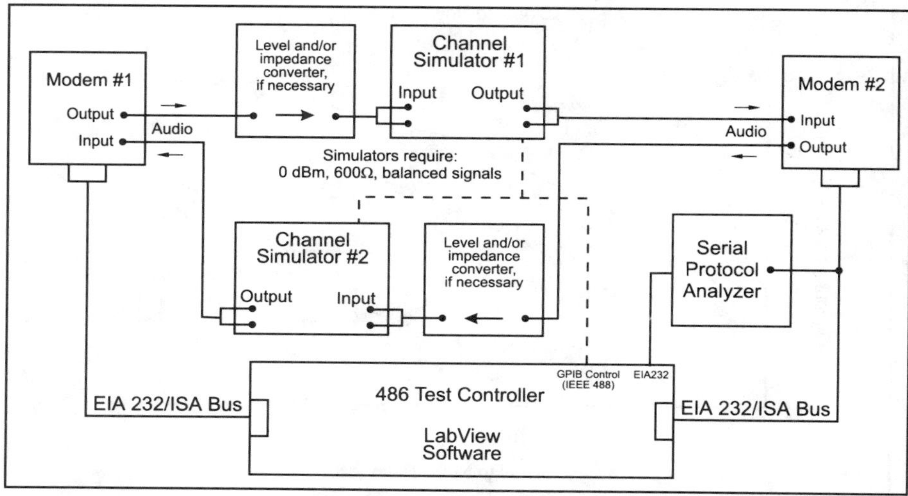

Figure 1—Block diagram of the HF modem test setup.

A *serial protocol analyzer* was used to monitor the serial port connection to the second (receiving) modem. A program running on the analyzer determined whether the test had timed out (no data had been transferred for a predetermined period of time). The test was controlled through the use of the National Instrument's program *LabView* (version 3.1.1), a *Windows*-based application capable of controlling, monitoring and measuring laboratory tests.

Two files were used to automate the test operation. The first was a test control file, containing a series of values representing the settings of the two simulators, the settings of the two modems and the file to be transmitted during each test. Results of the test were written to the second, a master log file.

For each test, *LabView* performed the following tasks:

1. Read the desired values of simulator settings, modem/protocol modes and parameters, and the file to be transmitted from the test control file.
2. Set the simulators to the desired values.
3. Initialized both modems to the desired mode.
4. Established a link between the two modems.
5. Started a timer.
6. Loaded the buffer of the transmitting modem with the test file.
7. Monitored the transmission and reception of the file.
8. Stopped the timer when the end-of-file was received.
9. Compared the received data to the original file to determine the success of the test.
10. Calculated the throughput and appended the test information to the master log file.
11. Flushed the modems' buffers in preparation for the next test.

LabView generated a front-panel display (Figure 2) that allowed the operator to monitor the progress of the test. This "panel" also supplied controls to permit the operator to run tests manually.

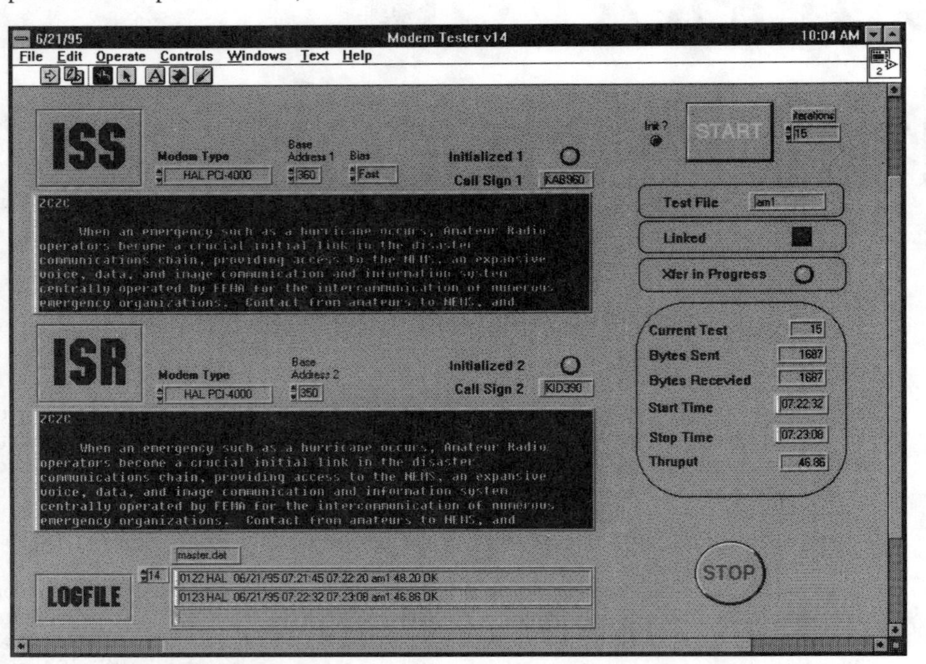

Figure 2—*LabView* generates this "control panel" on the PC monitor during testing.

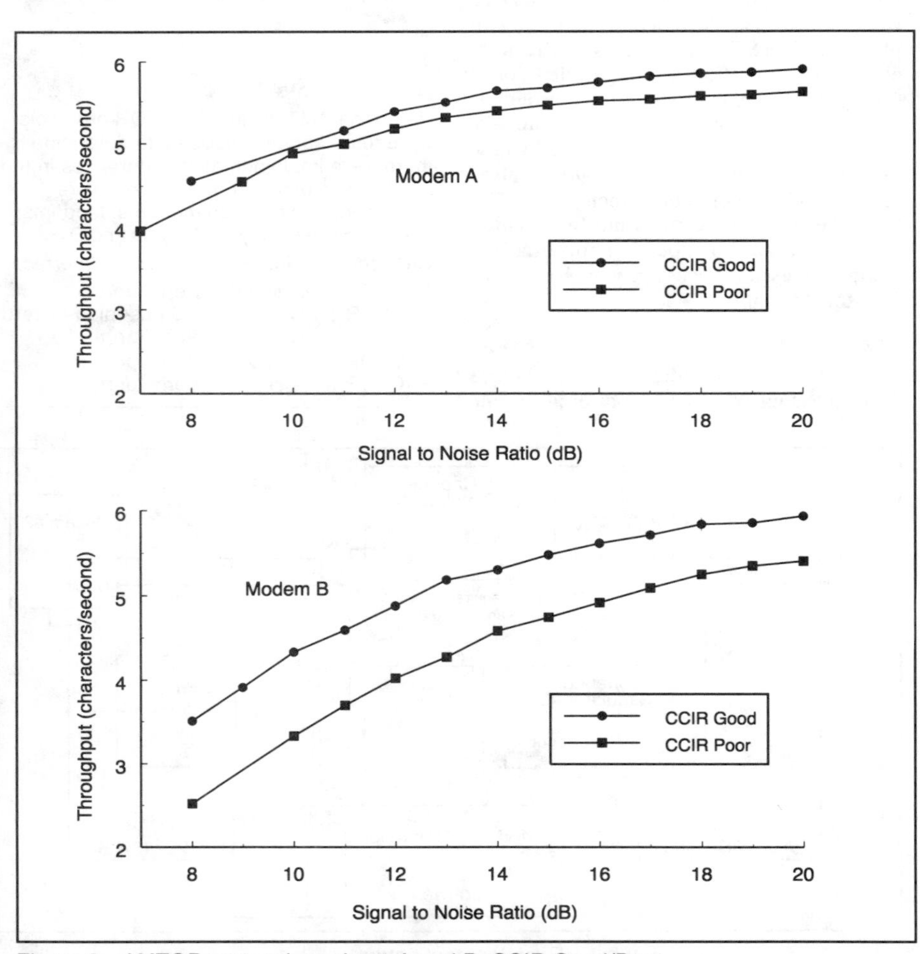

Figure 3—AMTOR protocol, modems A and B, CCIR Good/Poor

Which Parameters?

Because of the wide variety of protocols and the manner in which they are implemented, we didn't have enough time to test all possible parameter combinations. Because the goal of the tests was to determine the suitability of the modem/protocol combinations for use during emergency situations, we chose to test the protocols in their most robust mode. This meant testing only the ARQ mode. Because the ARQ mode is inherently error-free, there is no need to calculate error performance. Several of the protocols implemented algorithms for choosing modulation schemes. Otherwise, when manual settings were required, optimum settings were chosen based on the manufacturer's recommendations.

Seven or Eight Bits?

Because data can be transmitted in either 7 or 8-bit form, the type of test data used affects the throughput delivered by each protocol. Binary data and mixed-case ASCII data require a higher overhead when transmitted by a 7-bit protocol, thereby lowering the effective throughput. FEMA supplied a number of files it viewed as being "typical" traffic. These included both straight text and binary (word-processor format) content. To minimize the advantage 8-bit protocols have over 7-bit protocols, straight text files were chosen over binary files. When testing the AMTOR protocol, which transfers data as

7-bit characters using the Baudot character set, an all-uppercase version of the test file was used. For the other protocols, which use 8-bit characters, a mixed-case version of the same file was used.

Compression

All of the HF modems we tested used some form of data compression, ranging from real-time Huffman compression to optimized Lempel-Ziv compression performed prior to transmission. In keeping with ITS's goal of choosing an optimum mode, compression was enabled throughout testing. However, this causes a sensitivity to the size of the data file, as well as to the type of data the file contains. Huffman compression operates on a small amount of buffered data, and is relatively insensitive to file size and large data patterns. The CLOVER protocol uses compression software licensed from PKWARE, which operates on the data prior to transmission. The PKWARE software implements several compression schemes and, since it operates on the file in its entirety, can result in a more efficient and sophisticated compression. Therefore, it is more effective with larger files. In light of these differences, a large file of 15,183 bytes, which should favor neither compression method, was chosen from the selection supplied by FEMA. When compression is used, the data being transmitted between the two modems is no longer strictly text, but includes nontext, binary values, regardless of the file's original content.

Simulated Propagation Conditions

The selected propagation conditions, based on the Watterson model, were "CCIR Poor" and "CCIR Good" (see Table 2). The S/N ranged from 0 dB to +20 dB (to +40 dB for CLOVER). Because the HF channel simulators are band-limited to a 5-kHz bandwidth, this was the reference bandwidth for measuring the S/N as recorded in the log file.

Baud Rate

Baud rate was another factor in modem performance. The different baud rates used for each protocol are summarized in Table 3. Often the modems available for testing dictated what baud rate could be used with a particular protocol. Most protocols have a baud rate at which they are commonly used or are legally permitted to be used, while some protocols automatically vary the baud rate, depending on conditions (as reflected by the received error performance). Some also vary the baud rate as a function of the modulation scheme. In all cases, the highest baud rate is limited by FCC bandwidth restrictions. Tests were performed at these typical baud rates, or the protocol was allowed to automatically select the rate.

The throughput, while related to baud rate, does not linearly correspond to it. Variations in overhead sizes (due to the use of error-detection and correction schemes), repeated transmissions, and variations in the number of bits representing a symbol make significant differences in the throughput of a file or message. Commonly used units of throughput are *characters per second* and *words per minute*. The throughput, in characters per second, reflects the amount of time needed to transmit a file under the conditions of noise, multipath propagation and fading. The characters-per-second rate is used because this relates to the actual time needed to transmit a file, and it includes the transmission overhead. Only tests in which the data was received error-free were included in the results.

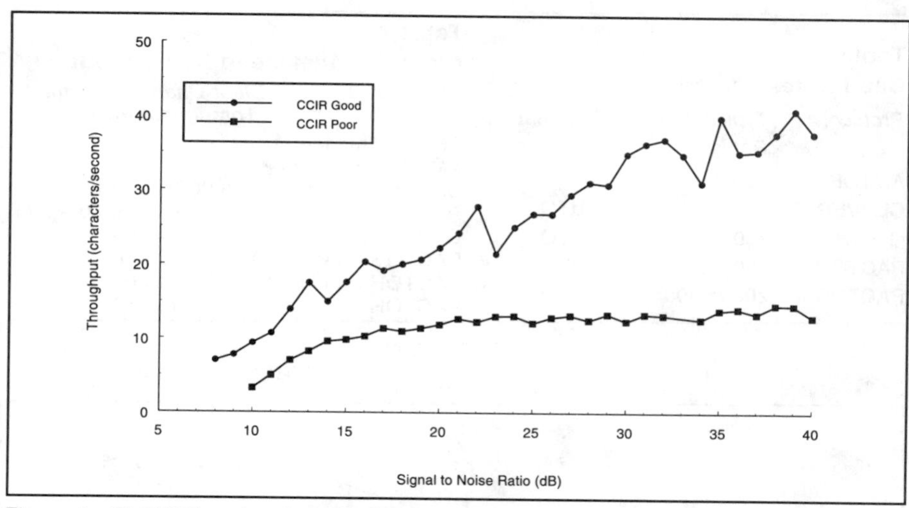

Figure 4—CLOVER protocol, CCIR Good/Poor

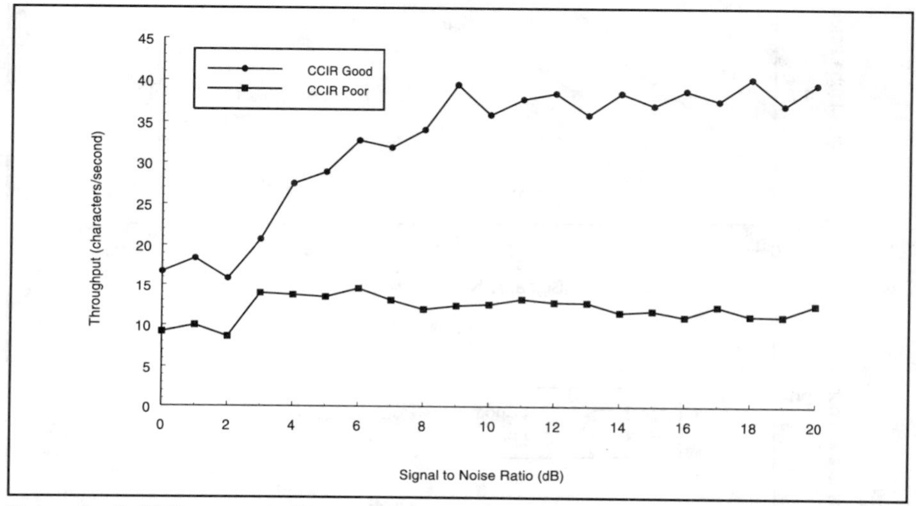

Figure 5—G-TOR protocol, CCIR Good/Poor

Table 2

Simulated Ionospheric Conditions Test Matrix for the Two-Path Watterson Model

Ionospheric Condition	*S/N (dB)*	*Fading Frequency Spread (Hz)*	*Multipath Delay (ms)*
CCIR Good	0 to +20 (to +40 for CLOVER)	0.1	0.5
CCIR Poor	0 to +20 (to +40 for CLOVER)	1.0	2.0

Test Limits and Time Constraints

Measurements over a clear channel (modems connected back-to-back without the simulators) were made to verify the modem setup and determine the optimal throughput. Table 4 presents the *clear-channel mean throughput* for each protocol and modem. Since implementations vary, one protocol can have significant differences in throughput between different models of modems. With this in mind, a protocol was tested on multiple modems whenever possible. This was the case with the AMTOR and PACTOR protocols only.

The clear-channel throughput was used to determine the limits of testing. While the ARQ mode will theoretically pass data error-free, it also can require an unlimited number

Table 3
Baud Rates Tested

Protocol	*Typical Baud Rates*	*Test Baud Rates*
AMTOR	100	100
CLOVER	92 to ~550	AUTO
G-TOR	100, 200, 300	AUTO
PACTOR	100, 200	200
PACTOR II	200 to 800	AUTO

Table 4
Clear Channel Mean Throughput

Protocol	*Clear Channel (optimal) Throughput (char/sec)*
AMTOR (modem A)	6.13
AMTOR (modem B)	6.26
CLOVER	45.60 (robust bias), 56.87 (normal), 69.05 (fast)
G-TOR	44.69
PACTOR (modem A)	25.46
PACTOR (modem B)	15.67
PACTOR II	133.53

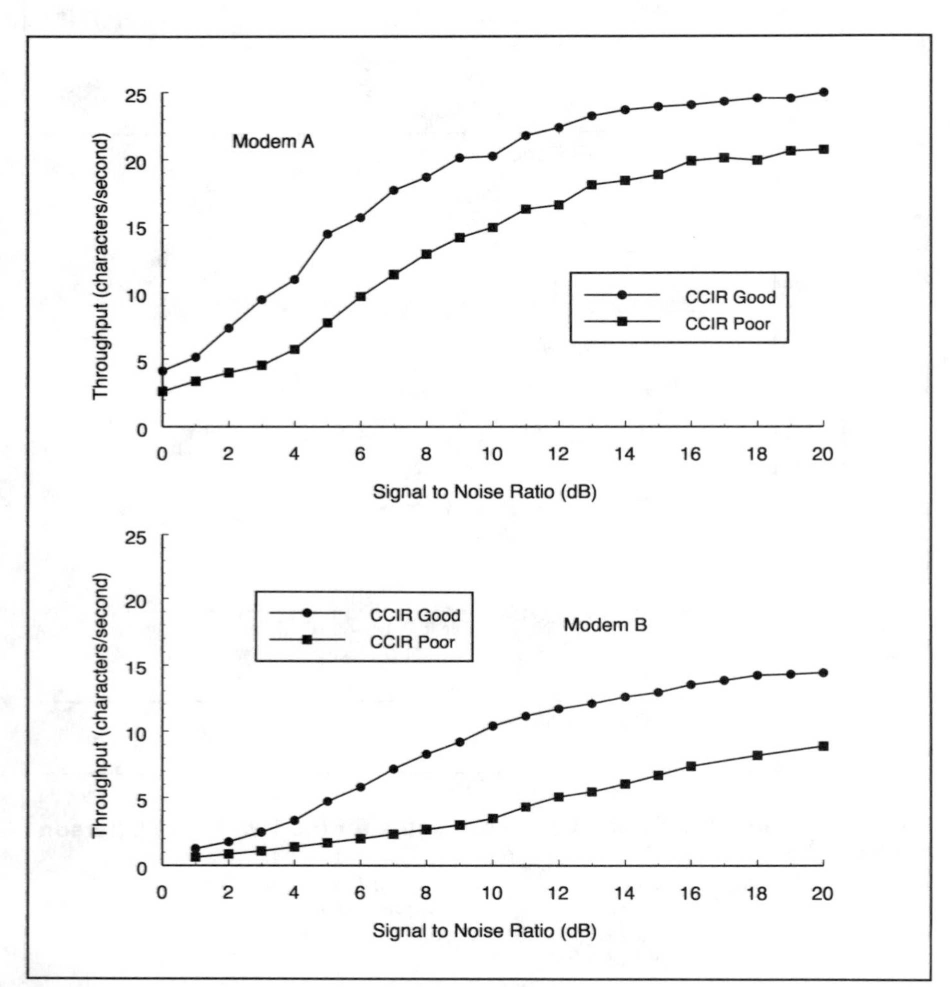

Figure 6—PACTOR protocol, modems A and B, CCIR Good/Poor

of retransmissions during poor signal conditions. Even under moderate conditions, occasional retries and associated handshaking will extend transmission time. As the transmission channel is degraded, the test time (time to pass a single data file) lengthens. At some point, the transmission time (and throughput) drops below a usable, practical level. Due to the limited time available to perform all of the desired tests, some breakpoint had to be set to restrict unnecessary testing. We decided to end testing when the throughput had dropped below 10% of its clear-channel value. While lower throughput may be usable in nonemergency situations, it is of little value during emergency situations, when rapid and reliable transmission of information is critical.

For an example of how time-consuming these tests are, consider the worst case: the AMTOR protocol. A suite of tests consists of 126 file transfers (21 transfers covering S/Ns of 0 dB to +20 dB in 1-dB increments, for two CCIR conditions, with three attempts each). At best, a throughput of six characters per second needed 42 minutes to transfer a 15,183-byte file, but as the S/N decreased, the transfer time increased. *Six days* were required to complete the AMTOR protocol suite of tests! This does not include the time needed to develop the drivers and initialization software necessary for each protocol/modem combination.

Results

The data collected during these tests is summarized in the plots in Figures 3 through 7. All plots display values of throughput (in characters/second) versus S/N (in dB).

The plots show the mean value of throughput for each S/N measured. The goal was to transmit a file three times for each S/N value. During development of the test procedure, more than three passes were made at certain settings. However, at low S/Ns, less than three passes are included in the plots due to errors and aborted tests. Again, the lack of time precluded us from running a test until three good completions were accomplished.

Over-the-Air Test Comparison

In 1993, FEMA ran a series of over-the-air tests as part of their program to evaluate new HF radio products.[3,4] Since FEMA and ITS tested two of the same modems and three of the same protocols, an attempt can be made to correlate the results of both tests.

FEMA chose frequencies characterized as either good or poor, based upon the *link quality analysis* (LQA) score supplied by an Automatic Link Establishment (ALE) transceiver, when such a score was available. When not available, RF output power was adjusted to simulate appropriate channel conditions. They estimated that a 10-dB S/N was the transition point between good and poor conditions. Tests were conducted across a network of three nodes at FEMA facilities in Berryville, Virginia; Denver, Colorado; and Urbana, Illinois; on frequencies authorized for FEMA use. A comparison of the FEMA tests to ITS's laboratory tests is shown in Table 5.

The LQA-based channel quality designators of "Good" and "Poor" do *not* correspond to the CCIR definitions for path conditions. Consequently, you can't make a direct comparison of data. Because the 10-dB breakpoint between LQA Good and LQA Poor is only an estimate, it's difficult to split up the laboratory measurement to approximate the LQA conditions. Therefore, a generalized comparison between the range of throughputs measured in the lab to the numbers obtained in the field is made, subject to these caveats.

The SITOR protocol was not tested in the lab because SITOR is a military version of the AMTOR protocol. While some differences in the two protocols may exist, ITS decided to drop SITOR from the tests, again due to time constraints. However, the SITOR field results are reasonably close to the AMTOR lab results. PACTOR II had not been introduced at the time of the 1993 tests.

In general, higher throughputs were recorded in the laboratory tests than in the field tests. The laboratory tests allowed for very high S/Ns that would reflect optimum (if not unrealistic!) real-world conditions and were not affected by the rapidly changing ionospheric conditions that challenge real HF

Table 5

FEMA Over-the-air Performance Test Results

Protocol/mode	*Compressed Data*	*File size (bytes)*	*Channel Quality*	*Over-the-air Throughput (char/sec)*	*ITS Lab Throughput (char/sec)*
SITOR ARQ	No	650	Good	4.42	3.9-5.9
			Poor	4.27	
PACTOR ARQ	No	650	Good	9.70	Not measured
			Poor	5.70	
PACTOR ARQ	Yes	650	Good	8.80	2.5-25.0
			Poor	Not measured	
CLOVER ARQ	Yes	3238	Good	28.03	7.0-40.0
			Poor	16.83	

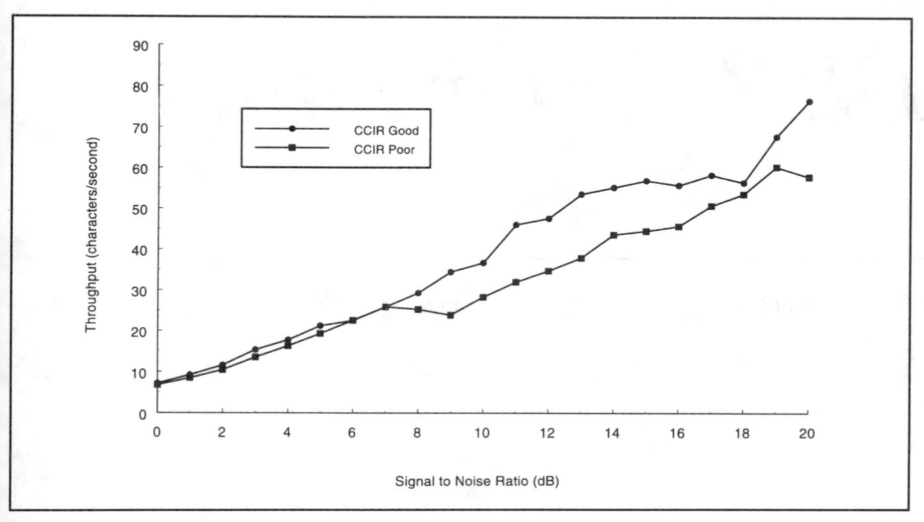

Figure 7—PACTOR II protocol, CCIR Good/Poor

operations. However, for providing a useful performance baseline of protocol performance, the laboratory tests are preferred. They can be repeated at any time of day or year, ensuring that the conditions will be the same for all tests.

Conclusions

In the case of CLOVER, the results show the obvious effects of phase modulation, automatic modulation selection, and the use of a sophisticated compression scheme. Throughput is notably dependent on file size and, although not tested, content. Compared to its clear-channel performance, the use of phase modulation demonstrates a susceptibility to multipath distortion. As the throughput varies, CLOVER changes phase modulation schemes (BPSM at low rates, through QPSM, 8PSM, 8P2A, 16PSM, to 16P4A at high rates) which shows up as points of discontinuity in the plotted data, notably at the 12, 23 and 35-character/second points.

CLOVER also uses a Reed-Solomon error-correction algorithm, the efficiency of which can be set to three levels (*Robust*, where 60% of the transmitted block contains data; *Normal*, 75%; and *Fast*, 90%). The Fast bias setting, which produced the highest throughput, was used during testing. Based on the test data, CLOVER seems to be best suited for transmitting large files. Note that the CLOVER protocol was tested over a S/N range of 0 to +40 dB, at the insistence of the manufacturer. All of the other protocols were tested over a range of 0 to +20 dB.

Similar results appear for the PACTOR II protocol, which also uses phase modulation, although the selection of phase level (DBPSK, DQPSK, 8-DPSK and 16-DPSK) is done manually, and not automatically as with the CLOVER protocol. For testing, the highest transfer rate (corresponding to the highest phase level modulation [16-DPSK]) was chosen. In both CLOVER and PACTOR II, the highest throughput measured under CCIR Good conditions is nearly half of the clear-channel throughput, and rolls off quicker for decreasing S/Ns than does the throughput of protocols that do not use phase modulation schemes. However, the throughput of these two protocols is an order of magnitude greater than the "basic" AMTOR protocol.

An attempt was made to determine if the throughput of a particular protocol was specific to the manufacturer's implementation, or whether it was representative of all modems using that protocol. Of the five protocols tested, PACTOR and AMTOR are the most popular among amateur HF digital operators. CLOVER and G-TOR are proprietary protocols implemented by their respective developers only. PACTOR II is beginning to appear in modems produced by a couple of manufacturers. But at the time these tests were performed, only one manufacturer was shipping units. Consequently, multiple testing was performed only on the PACTOR and AMTOR protocols.

The throughputs of PACTOR and AMTOR were measured on two modems from different manufacturers. As can be seen in the plots in Figures 3 and 6, there is a difference in the two implementations of each protocol. In the case of PACTOR, the difference is significant; the throughput of modem A is nearly twice that of modem B even though the test conditions were identical for both modems. No attempt was made to determine why this difference existed, but it raises questions regarding interoperability of modems from different manufacturers using a common protocol.

In testing the two implementations of the AMTOR protocols, functional differences emerged during the test development phase. One modem appeared to transmit only those characters contained in the BAUDOT character set. Undefined characters were received as a question mark. The other modem could transmit additional (punctuation) characters, apparently using the character positions the BAUDOT set listed as undefined. This could cause problems when trying to communicate between these two modems over a mixed-modem link. While AMTOR is not an appropriate protocol for use in transferring binary data or heavily formatted text, the incorrect transfer of certain punctuation characters could corrupt the meaning of a text message.

Obviously, more extensive testing must be done before a single protocol can be recommended for use by FEMA. Before that is done, the overall requirements, including the range of channel conditions and the characteristics of the traffic (content, size and frequency of transmission), must be better defined.

Notes

[1] No doubt you'll notice that HF packet is missing from the list. Although it is popular on VHF and above, packet is often a poor choice for HF communication. This protocol is not particularly robust. It requires excellent signals at both ends of the path and also stable conditions for reasonable efficiency. With this in mind, HF packet was eliminated from consideration as a "serious" contender.

[2] Two narrowband, Watterson model HF propagation channel simulators created the ionospheric propagation conditions for each test. Two degraded conditions were used: CCIR *Good* and CCIR *Poor*.

[3] H. F. Wetzel, "HF Modem Over-the-Air Performance Test Report, Globe Link Corporation HF Modem Model GLC-1000A," Federal Emergency Management Agency, Washington, DC.

[4] H. F. Wetzel, "HF Modem Over-the-Air Performance Test Report, HAL Communications PCI-4000 HF Modem," Federal Emergency Management Agency, Washington, DC.

Dennis Bodson, W4PWF, ARRL Roanoke Division Vice Director, is employed by the National Communications System, 701 South Court House Rd, Arlington, VA 22204-2198. Tim Riley, Stephen Rieman and Teresa Sparkman are employed at the Institute for Telecommunication Sciences, 325 Broadway, Boulder, CO 80303.

QST

Measuring 9600-Baud Radio BER Performance

DSP techniques make testing a G3RUH-compatible radio easy.

By Jon Bloom, KE3Z

One of the jobs of the ARRL Lab is to test the performance of equipment sold to amateurs. With the new crop of 9600-baud radios coming out, we had to develop a technique for testing their performance. The best way to test the performance of a radio used for digital communication is, by far, to test the bit-error rate (BER) that the radio provides under various conditions. BER is a conceptually simple metric that answers the question: How many of the bits get through correctly when a data stream is passed through the system?

The system we developed uses a Texas Instruments DSP Starter Kit (DSK) board that includes a TMS320C26 processor. Figs 1 and 2 show the circuitry of the BER test box built in the ARRL Lab. The DSP generates a test signal that is passed through the system under test. For receiver testing, the signal modulates a low-noise FM signal generator that feeds the radio being tested. The demodulated output of the radio goes to the DSP input so that the demodulated signal can be compared to the transmitted signal. To test transmitters, the DSP output test signal is applied to the modulation input of the transmitter under test. The transmitter RF output is attenuated to a low level, then applied to the Lab-built test box where it is mixed with an unmodulated signal from a signal generator. The resulting IF signal is demodulated by a low-distortion demodulator, and the demodulated signal is routed to the DSP input for comparison with the generated signal.

225 Main Street
Newington, CT 06111
email: jbloom@arrl.org

G3RUH Signals

The 9600-baud system used for amateur packet radio, both terrestrially and via the UoSat packet satellites, uses signals handled by the G3RUH modem design. In this system, the binary data stream coming out of a TNC is first "scrambled" to remove any dc component of the signal. Scrambling, which has nothing to do with encryption or data hiding, is simply encoding that ensures that, on average, there are as many 1-bits in the data stream as there are 0-bits. The average voltage of the data signal is thus constant. The scrambled data signal is then used to generate shaped pulses. The shaped-pulse signal is what is applied to the modulation input of the FM transmitter.

On reception, the shaped-pulse

signal is filtered and limited, to recover the scrambled data stream. A clock-recovery circuit generates a clock signal that is synchronous with the incoming data. Using this clock, the modem descrambles the data stream, with the end result being a binary data signal identical to the transmitted data signal.

Since our test system has to generate and receive signals like those of the G3RUH modem, some discussion of the pulse shaping used is necessary. The general problem to be solved when sending digital data using FSK is to limit the bandwidth of the baseband (data) signal before applying it to the modulator. But you have to be careful how you do that.

When we limit the bandwidth of a pulsed signal, we necessarily stretch the pulses in time. That is, a signal that is bandlimited cannot also be time limited. That means that in limiting the bandwidth, we cause each pulse to overlap adjacent pulses. In theory, each pulse overlaps *all* of the other pulses, but as you go further away in time from a particular pulse, its amplitude gets smaller and smaller.

So, we need to find some way of allowing the pulses to overlap their neighbors without interference. One approach to doing so is to shape the

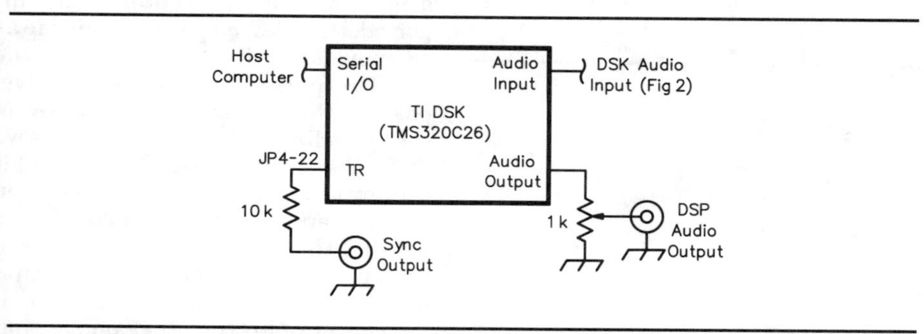

Fig 1—Diagram of the BER test connections to the TI DSK.

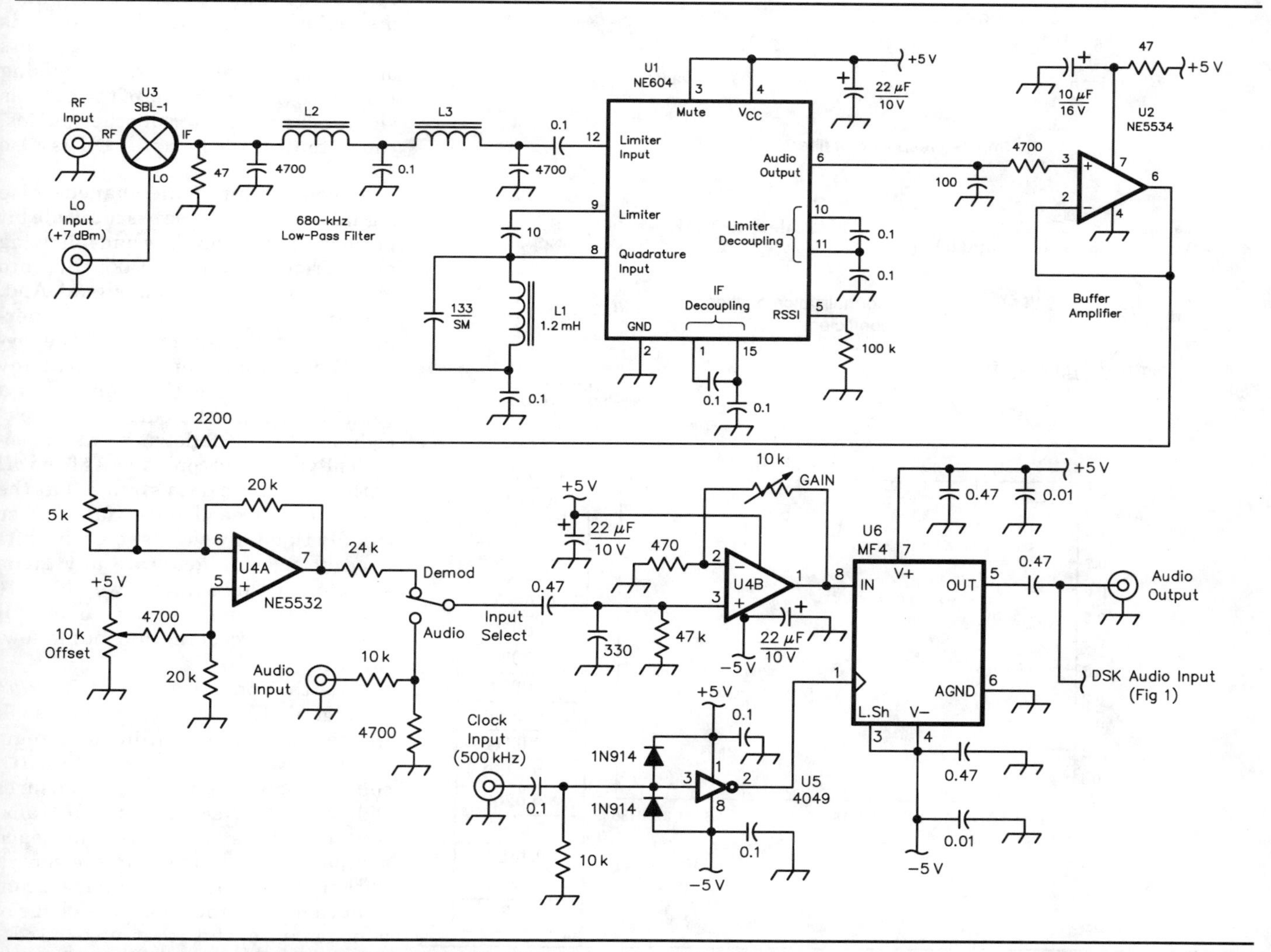

Fig 2—Schematic diagram of the BER test box circuit.

L1—1.2-mH Toko 10RB fixed inductor. (Digi-Key part TK4401-ND.)

L2, L3—41 turns #28 enam wire on a T-50-1 toroid core.

pulses so that, while each pulse does overlap its neighbors, the amplitude of that pulse is zero at the center of each of the other pulses. That way, we can sample the signal at the center of each pulse period and see only the signal from the current pulse; none of the other pulses contributes any amplitude to the signal at that time.

The spectrum of one pulse that has these characteristics is straightforward. It is flat from 0 Hz out to some chosen frequency, then rolls off with a cosine-shaped curve, reaching –6 dB at one-half the baud rate.[1] It continues rolling off with this cosine curve until it reaches zero. In the G3RUH system, the spectrum begins rolling off at 2400 Hz, reaches the –6-dB point at 4800 Hz and reaches zero at 7200 Hz.

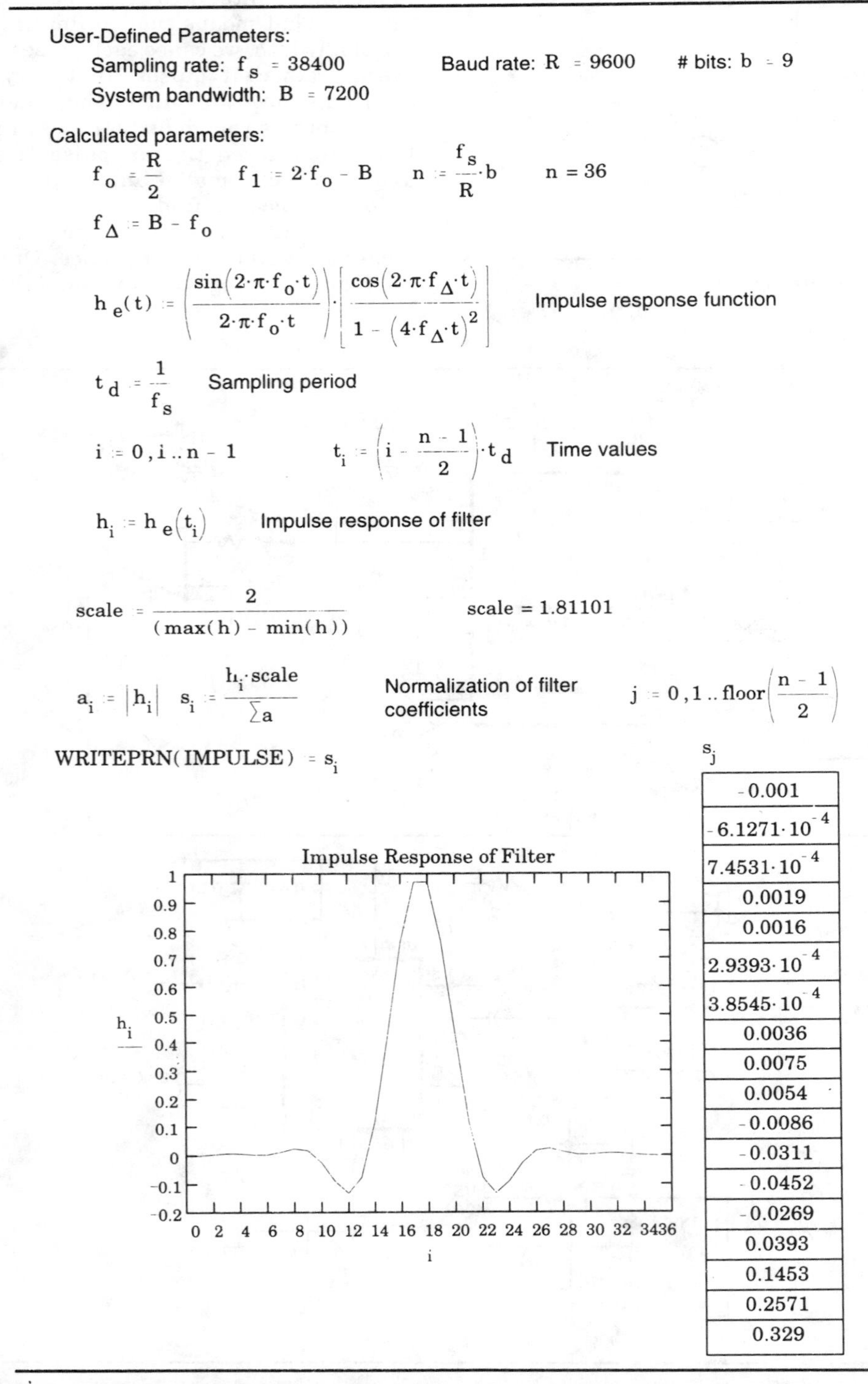

User-Defined Parameters:

Sampling rate: $f_s := 38400$ Baud rate: $R := 9600$ # bits: $b := 9$

System bandwidth: $B := 7200$

Calculated parameters:

$$f_o := \frac{R}{2} \qquad f_1 := 2 \cdot f_o - B \qquad n := \frac{f_s}{R} \cdot b \qquad n = 36$$

$$f_\Delta := B - f_o$$

$$h_e(t) := \left(\frac{\sin(2 \cdot \pi \cdot f_o \cdot t)}{2 \cdot \pi \cdot f_o \cdot t} \right) \cdot \left[\frac{\cos(2 \cdot \pi \cdot f_\Delta \cdot t)}{1 - (4 \cdot f_\Delta \cdot t)^2} \right]$$ Impulse response function

$$t_d := \frac{1}{f_s}$$ Sampling period

$$i := 0, 1 .. n - 1 \qquad t_i := \left(i - \frac{n-1}{2} \right) \cdot t_d$$ Time values

$$h_i := h_e(t_i)$$ Impulse response of filter

$$scale := \frac{2}{(\max(h) - \min(h))} \qquad scale = 1.81101$$

$$a_i := |h_i| \qquad s_i := \frac{h_i \cdot scale}{\sum a}$$ Normalization of filter coefficients $$j := 0, 1 .. \mathrm{floor}\left(\frac{n-1}{2} \right)$$

$$\mathrm{WRITEPRN(IMPULSE)} := s_i$$

s_j
-0.001
$-6.1271 \cdot 10^{-4}$
$7.4531 \cdot 10^{-4}$
0.0019
0.0016
$2.9393 \cdot 10^{-4}$
$3.8545 \cdot 10^{-4}$
0.0036
0.0075
0.0054
-0.0086
-0.0311
-0.0452
-0.0269
0.0393
0.1453
0.2571
0.329

Fig 3—This *Mathcad* worksheet calculates the impulse response of the FIR filter used to generate the shaped-pulse signal.

In the time domain, the pulse shape that results from such a spectrum has a maximum at the center of the bit period, then decreases in amplitude as we move away from the center of the bit. The pulse signal goes negative, passing through zero at the center of the preceding and following bits. As we go farther away in time from the bit center, the signal alternates between positive and negative values, always passing through zero at the center of each bit. There are other spectra that have pulse shapes that reach zero at the center of all the other pulses, but the benefit of the raised-cosine spectrum is that the pulse amplitude falls off rapidly with time. This is important because any amplitude or phase distortion present in the system is likely to cause the zero-crossing points of the pulse to shift in time, causing ISI. Since the amplitude of the pulse is small near the center of other pulses, the potential for harmful ISI is also small.

Since each bit of the shaped-pulse signal now extends across multiple bit periods—both preceding and following bits periods—we must take this into account in generating our signal. And, since this is DSP, what we are generating is a *sampled* version of the signal. What we must end up with at any sample is a signal that comprises a component from the current bit, preceding bits and following bits. Theoretically, we need components from *all* of the bits in the data stream, but the amplitude of each pulse falls off so rapidly that only a few successive bits need be used to generate any given sample. In this system, we chose to include components of the current bit and the four preceding and four following bits.

Note that only 1-bits contribute to the signal; 0-bits generate no pulse. If we were to send a continuous stream of 0-bits, we'd get no pulses at all. (Of course, the scrambler circuit ensures that will never occur in a real transmitted data stream.) So, for each sample we need to add up the contribution of the current bit and the contributions of eight other bits, some of which may be 0-bits that make no contribution. The contribution of a particular preceding or following 1-bit is he amplitude of its pulse at the resent time. To keep things simple, e use a sampling rate that is a multiple of the baud rate—and thus a multiple of the pulse rate.

The method used to implement this ulse forming is an FIR filter. The mpulse response of the filter is imply the samples that comprise a ingle shaped pulse, extending over ine bit periods. Our sampling rate is our times the bit rate, so the impulse esponse is 4×9=36 samples long. A *Mathcad 5.0* worksheet, shown in Fig , calculates the impulse response.

If we feed a single sample of amplitude 1 into the filter, preceded and ollowed by 0-amplitude samples, the esulting output will be a single copy f our shaped pulse, as shown in the *Mathcad* graph in Fig 3. To generate ur data stream, we input the present ata value (1 or 0) at one sample, follow it with three samples of 0, then nput the next data value. At any time, he FIR filter contains zeroes in all xcept (possibly) 9 locations, representing the 9 successive data bits. The utput of the filter at any sample time omprises components of 9 data ulses, which is what we want for our haped-pulse signal.

Now that we know how to generate he shaped-pulse signal from a data tream, we need to think about how to enerate the data stream itself. We want to generate a data stream that nimics the scrambled signal of the G3RUH modem. The scrambler in the modem is a tapped shift register with feedback. The logic equation for this circuit is:

$y = x_0 \oplus x_{12} \oplus x_{17}$

where y is the output of the scrambler, x_0 is the current input bit, x_{12} is the 12th previous input bit and x_{17} is the 17th previous input bit. Generating the test data stream is done by making x_0 always a 1, implementing a shift register in software, and calculating y for each new data bit—once every four samples. Note that this is essentially the same signal generated by the G3RUH modem in its BER test mode.

Counting Bit Errors

Having generated a test signal, we now need to consider how to compare the signal coming back from the system under test to the signal we transmitted. In passing through the tested system, the signal will be delayed by some amount, and the amount of delay will vary from one system to another. What we need to do is determine what the system delay is, then remember what our transmitted signal was that far back in the past in order to compare it to the samples of the received signal. But we are only interested in the value of the received signal at one time during each bit period: the center of the bit.

The DSP software handles this need using a two-step delay process. First, the operator will tell the DSP system how many samples of delay there are between the transmitted and received signals. Of course, the system delay is not likely to be kind enough to let the center of the received bits fall right onto one of our samples—there are only four samples per bit, after all. So, the second step is to adjust the phase of the DSP's transmit and receive sample clocks. The combination of these two techniques lets us adjust the DSP's delay with fine resolution.

This calibration procedure is performed using a dual-channel oscilloscope. One channel is connected to the *sync* signal from the DSK, the other channel is used to display the demodulated signal that is being fed into the DSK input. The operator first commands the DSK to generate a calibration signal. The DSK does so by passing a single 1-bit through the FIR filter described previously, followed by 0-bits. The 1-bit is repeated every 72 samples, resulting in a single shaped pulse every 2 ms. The oscilloscope is triggered on the sync signal and set for 0.2 µs/div. The operator then sends commands to the DSK to alter the number of samples of delay between the transmitted and received signals. The sync signal, a short pulse, is output during the sample the DSK believes to be the center of the received bit. Thus the operator simply adjusts the delay value, which causes the received signal trace on the oscilloscope to move relative to the sync

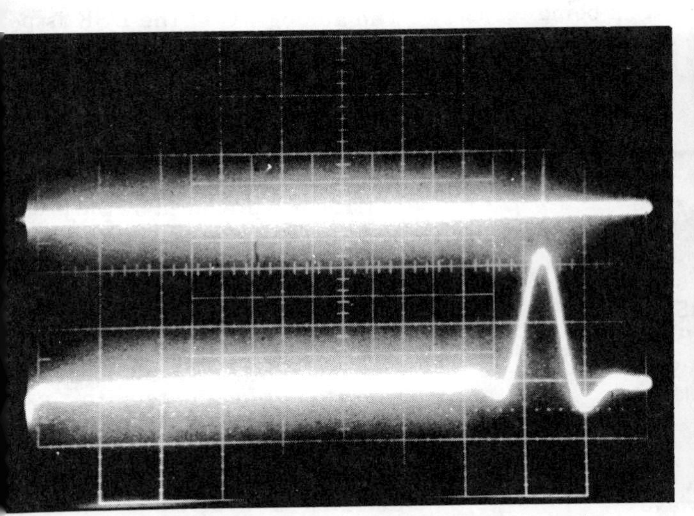

Fig 4—The first-phase calibration signal. When properly calibrated, the calibration signal (lower trace) is aligned with the sync pulse (upper trace) as shown. The oscilloscope is set for 0.2 µs/div. The sync pulse is very narrow and hard to see in this photograph but shows up well on the oscilloscope screen.

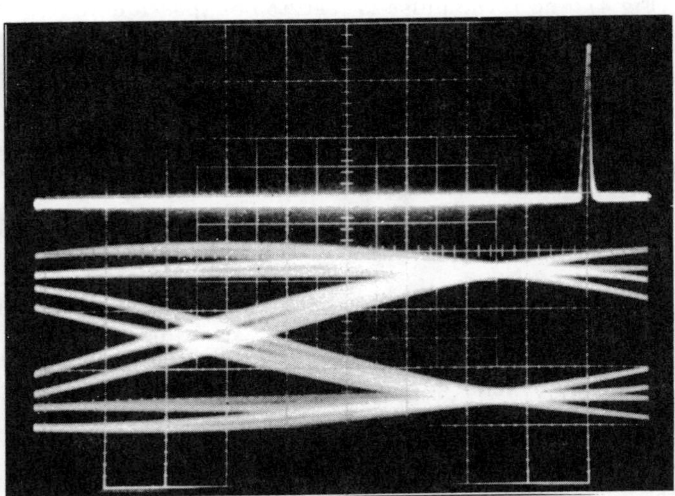

Fig 5—Second-phase calibration uses the BER test signal and sync pulse to produce an eye pattern. The oscilloscope is set for 10 µs/div. The center of the eye is adjusted to lead the sync pulse by 14 µs as shown.

pulse. When the sync pulse is aligned with the shaped pulse of the received signal (Fig 4), the DSK is at the proper sample-delay value.

To achieve final calibration, the DSK clock phases must be adjusted to take into account the part of the system delay that is less than one sample period long. This is done by commanding the DSK to output its test signal and switching the oscilloscope to 10 µs/div. Now the oscilloscope displays a sync pulse at the right side of the screen, along with the *eye pattern* of the received data (Fig 5). The proper sampling point of the bits is the point where the eye is most "open." However, there is a 14-µs delay in the DSK system between the actual sampling point and the sync pulse. Thus, the operator commands the DSK to step its clock phase until the center of the eye leads the sync pulse by 14 µs. This is not a hugely critical setting—a few-microsecond error isn't normally detectable in the measured BER. The delay does have to be in the ballpark, however.

There is one more detail to consider during calibration. Depending on the system being tested, the received signal may be inverted from the transmitted signal; the positive-going pulses we sent may now be negative-going. If this is the case, the DSK will count all the correctly received bits as bad and all the bad bits as correct. So, the DSK supports a user command to invert the sense of the received data. The polarity of the received signal is most easily seen during the first calibration phase (Fig 4), and if the pulse is seen to be inverted then, the DSK's "invert" command should be issued. That will cause no difference in the displayed signal, but the DSK will know "which way is up."

DSP Program Operation

The DSP program used to perform BER measurements is called DBERT. The object file, DBERT.DSK, is downloaded into the DSK from the host computer. Once DBERT is running, it communicates with the host computer via the DSK's serial port. The serial I/O communication mechanism was described in a previous article.[2] When serial I/O is occurring, the DSP interrupts must be disabled. This keeps the DSP from executing its signal-processing operations during serial I/O. For that reason, DSK operation is controlled by the host computer using a command-response sequence: The host sends a command to the DSK, the DSK executes that command—including any needed signal processing—and reports completion of the command back to the host, if needed. Each command to the DSK from the host is a single ASCII character. Some commands result in the DSK needing to send data back to the host. In that case, once the command is completed by the DSK, it sends the data back to the host via the serial interface. Table 1 lists the commands supported by the DBERT program.

Because the DSK has to stop its signal I/O while communicating with the host, there is a potential problem in taking a measurement. When the DSK begins generating the test signal, it is possible that the system being tested will exhibit a response to the start-up transient from the DSK that inval dates the result. So, the BER test sig nal is started and the DSK wait 20,000 samples (about a half-second before starting to sample the receive signal.

We wanted to be able to test at con sistent signal-to-noise ratios in orde to establish reference levels for com paring different radios to one anothe Since each unit will exhibit a uniqu sensitivity, we needed some way o adjusting the input power level to ge a fixed output signal-to-noise rati The solution was to include a SINA measurement function in the DBER program. When the DSK receives th SINAD command from the host, it ger erates a 1-kHz sine wave. About a hal second after it begins generating th sine wave, the SINAD measuremen software begins sampling the inpu signal. It then takes 8192 samples fo measurement. Each measured sampl is squared and added to a running sur (the *signal-plus-noise* value). Th samples are also run through a narro 1-kHz IIR notch filter that remove the 1-kHz test signal, leaving only th noise. The output samples from thi filter are also squared and added to (separate) running sum (the *nois* value). After 8192 samples have bee processed, the two 32-bit sum value are returned to the host computer. Th host can then calculate the SINAD b expressing the ratio of the signal-plus noise value to the noise value in dB. more detailed description of this tech nique was published in an earlier *QE* article.[3]

The analog I/O of the DSK is pe formed by a TLC32040 integrated ana log subsystem. This chip includes 14-bit D/A, 14-bit A/D, sample-cloc generator and input and output pro grammable switched-capacitor filter (SCF). The TLC32040 is driven by 10-MHz clock produced by the proces sor chip. The frequency of this clock in combination with the program mable dividers in the analog chip, de termines the available sampling rate The design of the BER test softwar calls for a 38,400 sample-per-secon (sps) rate. Unfortunately, this exac rate isn't possible with the 10-MH clock. The nearest available rate i approximately 38,461.5 sps. Thi translates to a 0.16% error in th speed of the test signals, which i negligible in the context of BER mea surements. However, this error is su ficient to make the test signal unread able by a G3RUH modem since th

Table 1—DBERT Commands

Command	*Return value*	*Description*
I	None	Generates test BER signal
C	None	Generates calibration signal
4	Error count (2 words)	Performs 10,000-bit BER test
5	Error count (2 words)	Performs 100,000-bit BER test
+	None	Increment sample delay
-	None	Decrement sample delay
A	None	Advance clock phase
0	None	Normal data polarity
1	None	Inverted data polarity
N	None	Generate 1-kHz sine wave
Q	None	Quiet (DSK output = 0 V)
S	SINAD value (4 words)	Take SINAD measurement
V	Input voltage (1 word)	Report input voltage value
T	None	Generate test signal (4800-Hz sine wave)

clock-recovery loop in that design has a very narrow lock range. I discovered this early on in the development of this system. To check that the problem was in fact the speed difference and not some problem with the signal I was generating, I temporarily removed the 40-MHz master oscillator on the DSK board—from which the 10-MHz clock is derived—and connected a signal generator to the clock input through a Schmitt trigger inverter. By setting the signal generator to generate exactly the clock frequency needed to get a 9600-baud signal, I found that the G3RUH modem was quite happy to consider my BER test signal a proper one.

Another problem with the TLC32040 is that the input SCF is a band-pass filter that normally cuts off at about 300 Hz. While the filter cut-off frequency is programmable, it can't be set anywhere near the very low frequency needed for this application. So, the DBERT program configures the TLC32040 to operate without its input filter. This requires that an external antialiasing filter be added. As shown in Fig 2, and described in more detail later, an SCF that cuts off at 10 kHz was added to the BER test box to fill this need.

The final characteristic of the TLC32040 to note is that it includes no internal sin*x*/*x* correction—not surprising in a chip with a programmable sample rate. Fortunately, with a sampling rate of 38400 Hz and signals of only up to 7200 Hz, the sin*x*/*x* roll-off at the upper end of the signal spectrum is only about ½ dB. Still, this seems worth correcting. The TLC32040 data sheet describes a first-order IIR filter that can be used to perform this correction. The *Mathcad* worksheet of Fig 6 shows the calculation of the coefficients of this sin*x*/*x* correction filter, which is implemented in the DBERT program. Fig 6 also shows the predicted output response both before and after the correction, as well as the estimated group delay of the correction filter.

As noted, the calibration procedure requires a sync pulse. The DSK has no uncommitted output signal lines usable for outputting a pulse, so DBERT outputs the sync pulse on the serial data output line. Since the sync pulse is less than a microsecond long, this won't usually bother the serial I/O chip of the host computer. (We did find one computer that was occasionally confused by the presence of the sync pulse.) The RS-232 signal is available on one of the DSK's expansion connectors. A 10-kΩ resistor connected to this point brings the sync signal out to the front panel of the BER test box. The presence of the resistor protects the RS-232 line from accidental shorting or application of another signal.

Host Application Software

We used two different programs. One, which we won't cover in detail here, manages the Lab's computer-controlled signal generator for stepped BER measurements at various signal levels and frequencies. The

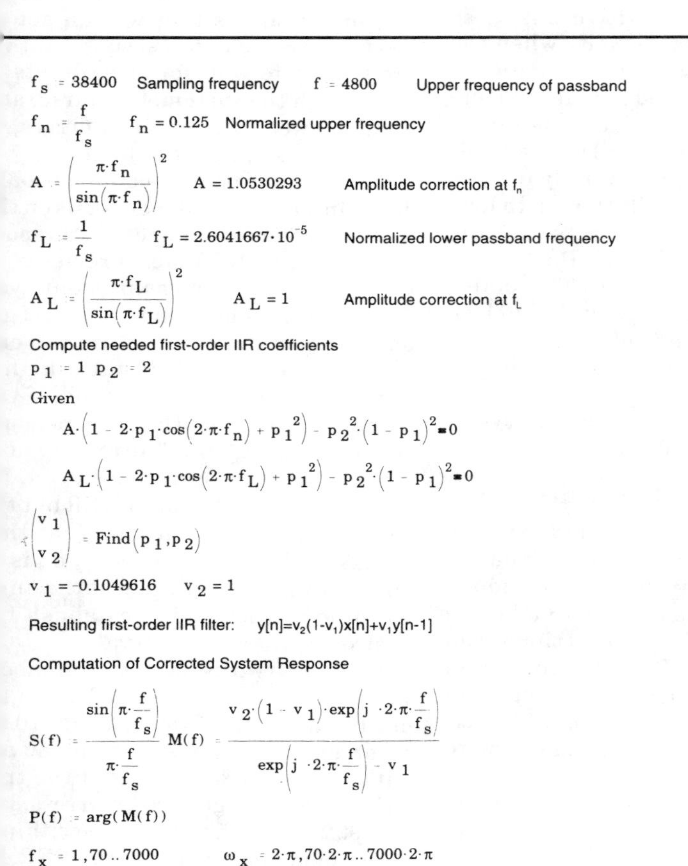

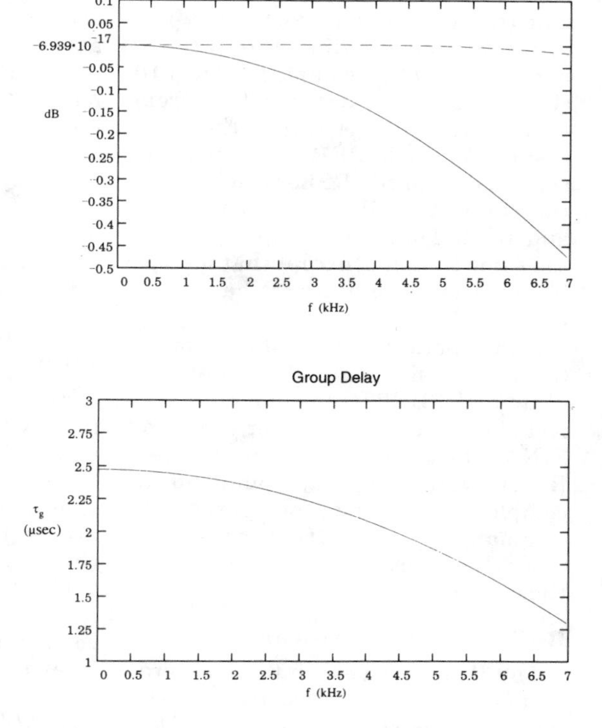

Fig 6—Correction of sin*x*/*x* roll-off is performed with a first-order IIR filter having the coefficients calculated by this *Mathcad* worksheet.

other program, BERT.EXE, is a C program that communicates with the DSK under operator control. Communication with the DSK is normally performed at 19,200 baud, although slower speeds can be used. To ensure that no serial overruns occur, BERT uses interrupt-driven serial I/O.

The BERT program is quite straightforward. It accepts keyboard commands from the operator and sends them to the DSK. The command set is the same as that of DBERT—BERT just passes these commands through to the DSK—with a couple of additions. Also, BERT "knows" about those DSK commands that generate a response from DBERT. When such a command is given, BERT waits for the response from the DSK, then converts the incoming serial bytes to binary values. In the case of BER test values (commands *4* and *5*), it prints out the reported number of errors and the BER, calculating the latter based on the number of bits the DSK was commanded to use in its test. For the SINAD command (*S*), BERT computes and prints the SINAD and the distortion percentage using the values returned by DBERT.

The DBERT program can perform BER tests using 10,000 samples or 100,000 samples. The BERT program also provides a 1-million sample BER test (the *6* command), which it performs by commanding the DSK to perform a 100,000-sample BER test 10 times. BERT sums the values from each of these tests to get the final result. As each BER test result is returned from the DSK, BERT prints the running total of samples, errors and BER. The *L* command performs the same function, except that it stops if 100 or more total errors have been reported.

If the operator selects the calibrate (*C*), idle (*I*), sine-wave (*N*) or quiet (*Q*) commands, BERT remembers the selected command. Whenever a BER or SINAD test command is performed, BERT waits for that command to complete, then sends the appropriate command to place DBERT in the most recently selected mode from among those listed.

BER Test Box Hardware

The BER test box contains several op-amp amplifiers used to keep the input signals within the range of the TLC32040 analog input. The box also contains a mixer and a demodulator, shown in Fig 2, designed by ARRL Lab Engineer Zack Lau, KH6CP. To test transmitters, the transmitter output is attenuated with a high-power attenuator down to a level that the SBL-1 mixer, U3, can handle. The LO input of the SBL-1 is driven by a +7-dBm signal from a signal generator set to a frequency 373-kHz below or above the transmitter frequency. The output from the SBL-1 passes through a low-pass filter that removes the sum frequency, leaving only the 373-kHz difference frequency. This signal is applied to an NE604 FM IF chip that contains a limiter, IF amplifier and quadrature FM demodulator. The frequency at which the demodulator operates is set by L1 and its associated silver-mica capacitor. When Zack built this circuit, the components he used just happened to fall at 373 kHz. If you reproduce the circuit, it's likely that your copy will work at a slightly different frequency. This shouldn't present a problem; the exact operating frequency isn't critical.

Since the NE604 operates from a single 5-V supply, its output is a positive voltage. This signal is fed into U4A, which amplifies the signal and also removes the positive offset, so that the result is zero volts when the input signal is at 373 kHz. U4B amplifies this demodulated signal or an external input signal (usually the output of a receiver being tested) with adjustable gain. The signal is finally filtered in an MF4 4th-order Butterworth low-pass SCF, U6, that acts as the antialiasing filter for the DSP input. The cut-off frequency of U6 is determined by its input clock, with the clock being 50 times the cutoff frequency. In our test set, this clock is supplied by an external function generator, but it could as easily be provided by a crystal oscillator and divider chain.

Interpreting BER Measurements

Bit errors are (or should be) due mainly to corruption of the signal by noise. Thus, they should be random. That being the case, if you make the same measurement several times, you are likely to get different results each time. But the more bits you send through the system, the more the effect of noise is averaged and the more consistent the measurements will be. That raises the question: How many bits do you need to use to get a valid result? The answer to that question depends on what you mean by "a valid result." The more bits you use, the closer your measurement will be to the "true" BER you would get if you sen an *infinite* number of bits through th system. Since sending an infini number of bits through the system is little, well, impractical, we need come up with some guidelines f selecting a finite number of bits to us

Unfortunately, in order to do that w have to make some assumptions abo the character of the noise in th system. Fortunately, the assumptio we make hold up pretty well for re systems. (There's nothing new abo this; we make assumptions about th character of system noise all the tim For example, when we relate noise system bandwidth we often assum that the noise is uniformly distribut across the spectrum of interest.)

What we find is that with a give number of bits in the sample set, ar a given number of errors within tho bits, we can establish a *confiden interval*. The confidence interval tel us how likely it is that our measur ment is within some specified factor the "true" BER. For example, we migh find that our measurement gives us a 80% confidence that the value within a factor of 3 of the true BE The end result is that we can get good a measurement as we want if w are willing to wait for enough bits go through the system. Of course, at low BER we don't get many errors, we need to send a *lot* of bits!

What's interesting about the conf dence interval is that it depends on th number of errors detected. Suppo you made two BER measurements. for the second measurement y double the number of bits sent and g double the number of errors, you en up with the same BER but a high confidence. On the other hand, if y double the number of bits but measu the same number of errors (you' measuring at a lower noise level, f instance), you get a lower BER but th same confidence interval. What th means is that all we need to do is ensure that we have at least the nu ber of errors needed to establish th desired confidence interval.

There are two useful sets of numbe we've used here in the ARRL Lab f our BER testing. If you measure 10 b errors, you are 95% sure that you a within a factor of about 2 of the tr BER. And if you get 100 bit errors, y are 99% sure that you are within factor of about 1.3 of the true BER.[4] you look at a curve of BER versus si nal-to-noise ratio, you'll find that factor-of-2 difference in BER occu with a fraction of a dB change in signal-to-noise ratio. That suggests that a measurement that results in 10 bit errors is a pretty good one, and a measurement that results in 100 bit errors is a *very* good measurement. That's why the *L* command exists in the BERT program. It pops 100,000 bits at a time through the system, stopping when 100 or more errors have been reported—because continuing to measure more bit errors is a waste of time—or when 1 million bits have been sent. Using a million bits means that a BER of 1×10^{-4} (100 bit errors) or worse can be measured with great accuracy, and a BER of 1×10^{-5} (10 bit errors) can be measured with decent accuracy. Of course, if you have time on your hands you can measure a BER of 1×10^{-6} by sending 10 million bits through the system. (That takes over 17 minutes at 9600 baud!)

Conclusion

The system described here has been used to measure a number of 9600-baud radios. Some of the test results will be presented in an upcoming *QST* article, and future *QST* reviews of 9600-baud radios will include measurements made using this system. The system has proven to be effective and easy to use. I should add that we spot-checked the results obtained with this system by measuring BER using a G3RUH modem. It gave BER results that were consistently slightly better because its input filter cuts off at a lower frequency than the SCF in the BER test box, improving the signal-to-noise ratio.

The software for this system, including the source code, is available for downloading from the ARRL BBS (203-666-0578) and via the Internet using anonymous FTP from ftp.cs.buffalo.edu. The file name is QEXBERT.ZIP.

Notes

[1]Couch, L. W., *Digital and Analog Communication Systems* (New York: Macmillan, 1993), p 179.

[2]Bloom, J., KE3Z, "Measuring System Response with DSP," *QEX*, February 1995, pp 11-23.

[3]Bloom, J., KE3Z, "Measuring SINAD Using DSP", *QEX*, June 1993, pp 9-13.

[4]Jeruchim, M. C., Balaban, P. and Shanmugan, K. S., *Simulation of Communication Systems* (New York: Plenum Press, 1992), pp 492-501.

The WA4DSY 56 KILOBAUD RF MODEM

A Major Redesign

By Dale A. Heatherington, WA4DSY

Abstract

In 1987 I designed a 56 kilobaud RF modem which was sold in kit form by GRAPES, the Georgia Radio Amateur Packet Enthusiast Society. This paper describes how the WA4DSY 56 kilobaud RF modem was radically redesigned to lower cost, reduce size, and improve reliability, manufacturability and useability. The reader is refered to the ARRL publication Proceedings of the 6th Computer Networking Conference, *page 68 for details on the original design.*

Overview

The original modem was implemented on 3 PC boards. It required both plus and minus 5 volts for the modem and 12 volts for the external transverter. The purchaser of the kit had to fabricate his own enclosure and obtain a suitable power supply. There were no status indicators. Only those hams with above average home brewing skills would attempt to build the unit. However, once built, the modem performance and reliability were quite good. Several high speed networks have been successfully built using these modems.

Unfortunately the large amount of skilled labor required to build a modem kit and to some extent, the cost have limited the wide spread adoption of these modems for high speed networking. I have redesigned the modem to address these issues.

The new design implements the modem on a single 4 layer printed circuit board powered from a 12 volt power supply. The PC board measures about 7 inches on each side. Signals produced by the new design are identical with the original and the new modems will interoperate with the old ones.

Most of the modem functions are implemented digitally in a Xilinx (tm) Field Programmable Gate Array (FPGA). The bandwidth limited MSK signal is generated digitally at 448 kHz to eliminate the need for analog double balanced modulators and a 90 degree phase shifter. This signal is up converted to the 10 meter band at 1 milliwatt. The converter is synthesized over a 2 mhz range (28-30 mhz). An external transverter converts the signal to the 222 or 430 mhz band. The delay between RTS and signal out is quite low, about 20 microseconds or 1 baud interval.

Note that this is a true *modem* which converts data to RF, unlike the G3RUH and K9NG designs which are baseband signal processors and don't do any *modulating* or *demodultating*.

The receiver is implemented with a single chip device and is synthesizer tuned from 28 to 30 MHz. A quadrature detector is used for FM demodulation. The demodulated signal is sliced using a circuit similar to the one in the original design which automatically adjusts the slicing level. The signal is then fed into the FPGA where clock recovery and data carrier detection are done digitally. The delay between receiving a signal and carrier detect indication is about 3 milliseconds.

The user interface has been greatly improved. Ten LEDs indicate received signal level. Other LEDs indicate *Request To Send*, *Data Carrier Detect* and *Ready*. The data interface is dual mode. A single switch selects CMOS or RS422 modes. The signals are presented on a DB15 connector wired to mate with the Ottawa PI2 Packet Interface Card. Other devices such as TNCs can be connected by wiring an appropriate cable and connectors.

Unlike the original design which allowed the user to reconfigure the modem for different baud rates, the data rate of the new design is fixed. Major changes are required to both the RF and FPGA circuits for use at any other baud rate. Part of the reason is because the first IF of 448 KHz must be 8 (or some power of 2) times the baud rate. Also, the receiver chip is being operated close to its maximum rate at 56KB.

Data Coding

All data coding is done in the FPGA chip. The chip is clocked at 14.31818 MHz. The transmit clock is obtained by dividing by 256. The exact baud rate is actually 55.9304 kilobaud, the same as the original design. The transmit clock signal is sent to the user on the RS422 interface. The user's transmit data source is expected to use the rising edge to clock out each bit. The modem samples data bits on the falling edge of the clock.

The transmit data stream is scrambled using the same shift register configuration as the original modem, a 17 bit register with feedback taps at stages 5 and 17. This is *not* compatible with K9NG and G3RUH 9600 bps modems. For more details on the scrambling system, see the section titled "Descrambler" below. After scrambling, the data enters the digital state machine where both NRZ to NRZI conversion and RF waveform table lookup operations are performed.

Modulator

This modem has no physical modulator. All RF waveforms are stored in EPROM. A digital state machine fetches the appropriate waveform segment from EPROM in response to the current data bit to be transmitted. 32 samples of the stored digitized waveform segment are read from the EPROM and sent to a Digital to Analog Converter (DAC) during each baud interval. The carrier frequency is exactly 8 times the baud rate to permit the splicing of different waveform segments together without phase discontinuities. The result is a 448 kHz bandwidth limited MSK signal. Unlike the previous version of this modem, there are no modulator related adjustment controls. The signal always has perfect phase shift and deviation characteristics. The signal is identical to the one produced by the original WA4DSY 56 KB modem.

Transmitted signal characteristics

- Modulation is MSK
- Bandwidth is 70 kHz at 26 DB down
- 3.5 DB amplitude variation
- 14 kHz FM Deviation
- 90 degree phase shift per baud

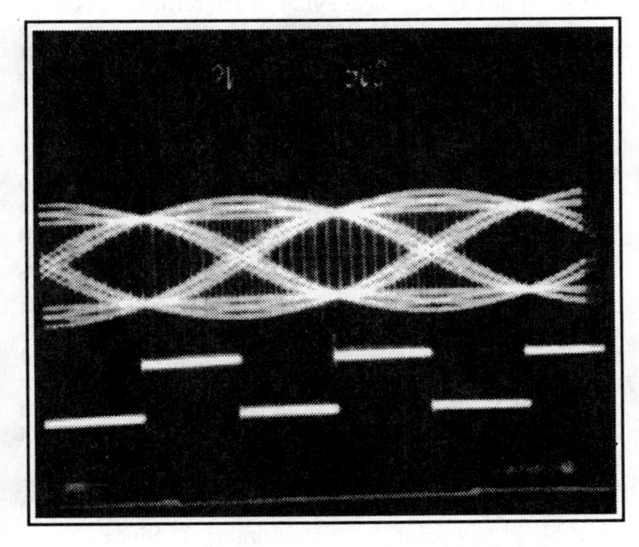

Top trace: Raw TX signal from 8 bit DAC

Lower Trace: 56KB TX Clock

Upconverter and IF

The 448 kHz MSK signal shown in the photo above is first filtered with an 80 kHz wide 3 section LC bandpass filter to remove digital sampling noise. It's then mixed with 10.245 MHz and converted to 10.693 MHz. The 10.693 MHz signal is passed through two 10.7 MHz (180 kHz BW) ceramic filters to remove the local oscillator and unwanted lower sideband (10.245 and 9.797 MHz). The undesired frequencies are reduced by at least 90 db.

The desired 10.693 MHz signal is then mixed with a VCO signal in the 39 MHz range. The lower sideband of the mixer output (28-30 MHz) is selected with a two section LC bandpass filter. Both conversions are done with NE602 frequency converter chips.

The 29 MHz signal is amplified 30 db by an MMIC chip and sent out to the users transverter on a BNC connector. The output level is adjustable from -10 DBM to +5 DBM.

The local oscillators are running at all times to assure instant response to the user's "Request to Send" control signal. The total delay is less than 20 microseconds from RTS to RF data signal out. Contrast this to the original design which required up to 6 milliseconds to start the transmitter.

Receiver

The receiver uses a Motorola MC13135 chip for all RF signal processing. The received signal in the 28-30 MHz range from the user supplied transverter is filtered by a two section bandpass filter before entering the receiver chip. The first local oscillator is a VCO in the 39-41 MHz range controlled by a synthesizer. The signal is mixed with the VCO to convert it down to 10.693 MHz. The signal is bandpass filtered with a single 180 kHz wide ceramic filter before being mixed with 10.245 MHz and converted down to 448 kHz. A 60 kHz wide LC bandpass filter provides both selectivity and deemphasis. Frequency modulation is recovered with a quadrature detector.

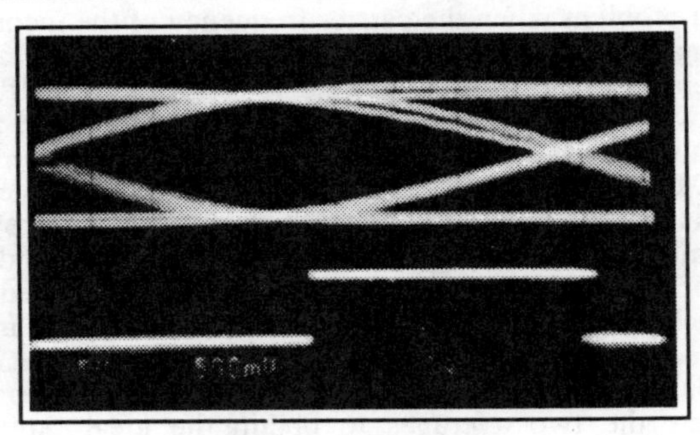

Receiver Eye pattern

Frequency Synthesizer

I used a Motorola MC145162 synthesizer chip for this design. It is programmed serially with a three wire interface. It has completely separate reference counters and divide by N counters for transmit and receive. The reference oscillator running at 10.245 MHz also drives both the receive down-converter and the transmit up-converter.

The VCOs are designed to cover 39 to 41 MHz, 10.693 MHz above the 29 MHz IF frequencies. The receive VCO is included in the MC13135 receiver chip. The transmit VCO is a Colpits transistor oscillator with an emitter follower output buffer.

Since there is no microprocessor in this modem, I needed a way to generate the data to program the synthesizer. I had the good fortune of having extra space in the EPROM; so I put the frequency data there. Up to 8 different bit patterns for independent TX and RX frequencies can be stored in the EPROM. The first thing the FPGA does after loading its configuration data is read one of the selected frequency bit strings into the synthesizer chip. Three switches code for the 8 frequencies. Any 8 frequencies can be programmed into the EPROM using a simple program written in C. The program will be supplied with the modem. If the user wants a custom set of frequencies, he must have access to an EPROM programmer or order a custom EPROM from the dealer. Still, this is superior to the original design which used custom crystals often requiring a 6 to 8 week wait and costing 10 to 15 dollars each.

Gated Tracking Data Slicer

Before the recovered signal can be used it must be processed to determine the state of the received bit, 1 or 0. This is done with an analog comparator chip. Its threshold is set exactly halfway between the voltage level of a "1" and a "0". It outputs a "1" if the input is higher than the threshold and a "0" if it's lower. There is a problem when the carrier frequency of the incoming signal changes. The voltage levels of the ones and zeros change so the threshold is no longer exactly half way between them. This causes an increase in errors. One common solution, which doesn't work very well, is to AC couple the output of the demodulator to the detector. This is fine if the short and long term average of the number of ones and zeros is equal. This ideal condition cannot be guaranteed even if a scrambler is used. A much better solution is to put some intelligence in the detector so that it averages the voltage level of the ones separately from the average of the zeros and then subtracts the two averages to obtain the ideal threshold level. This circuit doesn't care about the ratio of ones to zeros as long as there is a reasonable number of each. A scrambler is used to make sure there is a reasonable number of both ones and zeros. The circuit will compute the correct threshold if the input signal carrier frequency is anywhere within the expected range of the ones and zeros, in this case plus or minus 14 kHz. The data slicer used in this implementation is gated with the recovered clock. It only sees voltage levels near the center of a baud interval. A leaky sample and hold technique is used to grab the middle 1 microsecond of each bit. There is little variation in the peak levels from bit to bit thus reducing unwanted fluctuations in the slicing level.

Clock Recovery

Clock recovery is done digitally in the FPGA. There are no adjustments such as VCO center frequency as in the original design. The phase of a 3 modulus counter is compared with the data zero crossing times. The counter is driven by a 3.579545 MHz clock. The baud rate for each modulus is listed below.

Modulus	Baud Rate
• 63	56.818 (fast)
• 64	55.930 (on time)
• 65	55.069 (slow)

The counter can divide by 63, 64 or 65. The divide by 64 setting produces a 56 kHz clock. If the zero crossing was late relative to the counters terminal count then the counter is counting too fast. The counter modulus is set higher so it will be earlier next time. If the zero crossing is early the modulus is set lower so it will be later next time. If no zero crossing is detected the modulus is set to "on time" so the clock won't drift during strings of ones or zeros. This scheme only introduces about 0.5 microseconds of clock jitter (3%).

Data Carrier Detector

The data carrier detector is also implemented digitally in the FPGA. There are no adjustments. Two gates are used to separate data zero crossings which fall within plus or minus 12 1/2% (in sync) of the terminal count of the 3 modulus counter described above from the zero crossings which fall outside this range (out of sync). If the clock recovery circuit is phase locked, all zero crossings should fall within the 25% "in sync" window. This is true even at low signal to noise ratios. The "in sync" zero crossings cause a 5 bit counter to increment. The "out of sync" zero crossings cause the counter to decrement. The "carrier detected" flip flop is set when the counter reaches maximum count (31). The flip flop is reset when the counter reaches minimum count (0). The counter is designed not to overflow. It has "stops" at count 0 and 31. Carrier detect occurs when the clock recovery circuit has acquired phase lock and 31 more "in sync" zero crossings have occurred relative to "out of sync" zero crossings. This takes about 3 milliseconds or about half the time of the original 56KB modem with less falsing. Measurements show solid carrier detect even when the bit error rate is as high as 6%. Random noise can't assert carrier detect because the zero crossings have random timings and will occur with equal probably at any point in the baud interval. Since 75% of this interval is devoted to decrementing the 5 bit counter, it will quickly go to zero and reset the carrier detect flip flop. Periodic waveforms that are harmonically related to the 56 kHz clock frequency will trigger carrier detect if the clock recovery circuit phase locks to it.

NRZI to NRZ conversion

NRZ is a data signaling format in which zeros are represented by a certain voltage level and ones by another. NRZI is a signaling format in which zeros are represented by a change in voltage level while ones are indicated by no change. NRZI coded data is not affected by inverting the data voltage levels or the mark/space frequencies in the case of FSK. This modem converts the incoming NRZI data to NRZ data with a simple circuit consisting of a "D" Flip Flop and XOR gate. These components are in the FPGA chip.

Descrambler

A self synchronizing data scrambler was used in this modem for two reasons. First, it makes the data stream look like a random stream of ones and zeros regardless of the data being transmitted. This characteristic makes the tracking data slicer and clock recovery circuits work better. Second, it makes the RF spectrum look and sound like band limited white noise. In other words, the RF energy is spread evenly over the modems bandwidth and shows no single frequency lines regardless of the data being transmitted. Any potential interference to neighboring channels is limited to an increase in the noise floor instead of squeaks, squawks, and other obnoxious noises. This type of scrambling is also commonly used in high speed synchronous modems for telephone use.

The hardware to implement the scrambler and descrambler is very simple. It consists of a 17 bit shift register and two XOR gates, also known as a *Linear Feedback Shift Register* (LFSR). Each transmitted bit is the result of the exclusive ORing of the current data bit with the bits transmitted 5 and 17 bits times before. To descramble the data, it is only necessary to exclusive OR the current received bit with the previous 5th and 17th bits. If the data consist of all ones, the scrambler will produce a pseudorandom sequence of bits that will repeat after 131,071 clock pulses or every 2.34 seconds at 56 kilobaud.

This linear feedback shift register scrambling scheme does not violate the FCC prohibition against codes and ciphers because its purpose is to "facilitate communication" and the algorithm is publicly available.

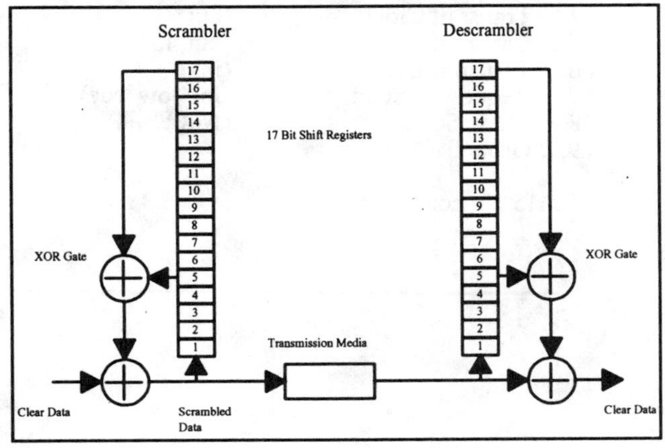

Scrambler Block Diagram

Note: G3RUH and K9NG scramblers use 17 bit shift registers tapped at stages 17 and 12. The sequence produced is not maximum length.

8 Bit FIFO and Bit Repeater Mode

To allow this modem to be used as a full duplex bit repeater I have included a first in-first out (FIFO) buffer and logic circuitry to route the received data bits back to the transmitter. When repeater mode is enabled, the data carrier detect signal will assert request to send. A 2 second watch dog timer prevents transmitter lockup. The FIFO buffer is 8 bits long. To allow for both plus and minus "bit slip", the FIFO does not start sending data until it's half full. The transmitter then pulls bits out of the FIFO at it's fixed clock rate while the receiver inserts bits into the FIFO at the receiver clock rate. The FIFO is reset when data carrier detect (DCD) drops. With a data rate difference between incoming and outgoing data of 0.01% a packet of 40,000 bits (5000 bytes) can be retransmitted before the FIFO buffer overflows. Since this is much larger than typical AX25 packets, I don't think this restriction will be a problem. Keep in mind that DCD must drop between packets to reset the FIFO.

A single internal switch enables the bit repeater mode. The external data I/O connector remains active to allow communication with a local computer.

Watchdog Timer

A 2 second watchdog timer is built in and may be bypassed with a switch setting. The timer is reset when RTS is false and begins timing when RTS is true. PTT will be turned off if RTS is not removed after about 2 seconds. A resistor and capacitor set the time-out value. This is the only analog timing circuit in the modem.

Remote Control

Computers used in amateur packet networks are often located with the TNCs, modems and radios in inaccessible places such as mountain tops. When the computer software crashes, which it often does, the control operator doesn't want to have to go to the distant site to push the reset button. A remote control reset function is built into this modem.

Normally open relay contacts are available for whatever use may come to mind. The contacts close when the modem decodes several milliseconds of a pseudorandom bit pattern sent to it from another modem in convenient reach of the control operator. A push button on the rear panel causes the modem to send the special sequence.

The sequence generator is an 8 bit linear feedback shift register with user specified taps. The tap locations are specified along with the frequency data in the EPROM. Each RX/TX frequency pair may have a different code assigned. Each set of taps produces a unique code. The receiving modem must see at least 256 bits of the pattern before it will start to close the relay. The sequence must continue for another several milliseconds to allow the relay to close. A single incorrect bit in the sequence will reset the decoder so that another correct sequence of 256 bits are needed to cause relay closure. The front panel "ready" LED will change from green to red when the code is being received.

Signal Level Display

Ten LEDs on the front panel indicate relative signal strength. The RSSI signal from the MC13135 receiver chip drives an LM3914 linear bar graph display controller chip. I have found this a most welcome feature. I've used it to map signal coverage areas by first setting my station to "ping" the local 56KB packet switch every 2 seconds; then with only a modem, transverter and antenna in the car, I can get a good idea of how well the packet switch covers various areas by watching the signal level LEDs and the DCD LED (and the road).

Tune-up and Test Aids

Since only half the EPROM storage was used for the main FPGA configuration code, state and waveform tables, I provided a switch which allows the modem to "boot up" using the other half of the EPROM. In the other half is an FPGA configuration for a direct digital frequency synthesizer which is used to sweep the modems tuned circuits. It sweeps a 200 KHz range centered on 448 KHz. A square wave is also generated on the receive clock output line for scope sync. The rising edge coincides with the 448 KHz center frequency. Adjustment of the filters for proper response shape requires only a 30 MHz dual trace scope. The scope must be adjusted so the rising edge of the "receive clock" square wave is centered and exactly one complete cycle is displayed. The other channel can then be used to probe various points in the modem to observe the frequency response envelope calibrated to 20 KHz per horizontal division. The receive filter can be checked if the an attenuator is placed between the TX and RX BNC connectors. The transmitter becomes the sweep generator.

A push button on the rear panel will activate the transmitter and send scrambled marks. The 2 second watchdog timer is automatically bypassed to allow transverter tune-up or power measurement.

Interfaces

Transverter: Power and PTT (Push To Talk) transverter signals are provided on a 5 pin DIN connector. The remote control relay contacts are also on this connector. BNC connectors are provided for the 29 MHz transverter IF signals. Pin assignments are as follows:

1. PTT
2. Relay contact
3 Ground
4. Relay contact
5. +12 volts @ 2 Amps

Power: The power input connector is a common 2.1 mm round DC power jack used in many other consumer electronic devices. Positive voltage is supplied on the center pin. A 12 volt 2.5 amp external switching power supply runs the modem and transverter.

Data: A female DB15 connector is used for the data interface. The pin assignments are the same as the Ottawa PI2 card. A single switch on the PC board changes the electrical standard from balanced RS422 to unbalanced CMOS. The RS422 interface is based on the 26LS32 and 26LS31 chips. The CMOS interface uses a 74HC244 chip.

Pin assignments

RS422

1.	No connection	
2.	+ Receive Clock	(out)
3.	+ Receive Data	(out)
4.	+ Transmit Clock	(out)
5.	+ Carrier detect	(out, low true)
6.	+ Transmit Data	(in)
7.	+ Request to Send	(in, low true)
8.	Mode Select	(Not Used)
9.	Ground	
10.	- Receive Clock	(out)
11.	- Receive Data	(out)
12.	- Transmit Clock	(out)
13.	- Carrier Detect	(out, high true)
14.	- Transmit Data	(in)
15.	- Request to Send	(in, high true)

CMOS

1.	No connection	
2.	Receive Clock	(out)
3.	Receive Data	(out)
4.	Transmit Clock	(out)
5.	Carrier detect	(out, low true)
6.	Transmit Data	(in)
7.	Request to Send	(in, low true)
8.	Mode Select	(Not Used)
9.	Ground	

10..15 No connection

Internal Option Switch Functions

Switch	Off	On
1.	RS422	CMOS
2.	Normal	RX Mute Disable
3.	Normal	Repeater Enable
4.	Normal	Scrambler Disable
5.	Normal	Key Transmitter
6.	Normal	Tune up
7.	Normal	Watchdog Disable
8.	Frequency Select 2	
9.	Frequency Select 1	
10.	Frequency Select 0	

Performance

Due to deadline and and other time constraints, I was unable to do a bit error rate test on the latest PC board revision. The previous version needed about 2 db more signal to achieve the same bit error rate as the original WA4DSY 56KB modem. This was a receiver problem related to excessive wideband digital noise getting into the receiver RF stages. This will be resolved before production.

Performance with off frequency signals seems to be at least as good as the original design, degrading only about 1 db with a 5 KHz frequency offset.

The response time from RTS to DCD has been measured at about 3 milliseconds. I'm using a TXDELAY value in NOS of 5 ms and haven't encountered any problems. This is much faster than the original design which required a TXDELAY setting of 15 ms.

The transmitter spectral bandwidth is about the same as the original design.

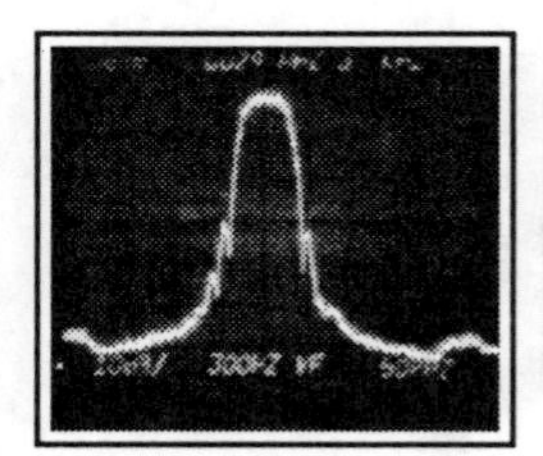

Transmited spectrum
Horizontal: 50 KHz/division
Vertical: 10 DB/division

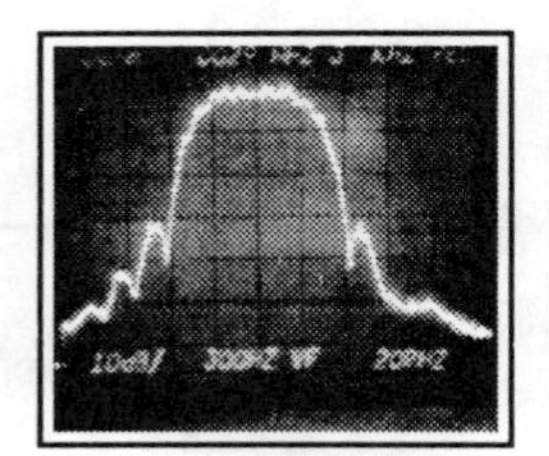

Transmited spectrum
Horizontal: 20 KHz/division
Vertical: 10 DB/division

Applications

This modem can find uses in several areas. When the original modem was introduced in 1987, the computing power available to the average ham was quite limited and had problems keeping up with 56KB data. Today (1995) the average ham can afford a 66 MHz 486 machine. Multitasking operating systems such as OS/2 and Linux running on these machines allow hams to set up their own TCPIP Web sites on the air. Applications such as a Web server are useless at 1200 baud. For this reason, I believe this modem has as much potential for use as a user LAN modem as it does for point-to-point links. Assuming the built in full duplex bit repeater works as expected, I hope to see many 56KB FDX user LANs spring up around the world. They would work just like an FM voice repeater except for the 70 KHz bandwidth requirment. There is one such LAN in Ottawa, Ontario, Canada.

We plan to put up a full duplex 56KB Metropolitan Area Network on 222.400 (input) and 223.85 MHz (output) here in the Atlanta, Georgia area using a 56KB modem, a receive converter, transverter and a Sinclair duplexer. Users, of course, only need a modem and transverter.

The U.S. now has a 1 MHz wide band (219 to 220 MHz) for "point-to-point fixed digital message forwarding". The band is divided into ten 100 KHz channels. This modem is ideal for that service.

Sales and Marketing

By now you're probably wondering how to get one of these modems. You can't, at least not right now. The modem is still in development. However, I'm negotiating with a well known manufacturer of packet radio equipment to produce and sell this modem. I hope to see them advertised for sale late in 1995 or early 1996. GRAPES will also be involved in modem sales.

A Prototype Modem

Editor's Note: They are now available from:

PacComm Packet Radio Systems Inc.
4413 North Hesperides Street
Tampa, FL 33614-7618
Phone: 800-486-7388 (Orders)
813-874-2980
Fax: 813-872-8696
Email: info@janrix.com
Web Site: http://www.janrix.com/paccomm/

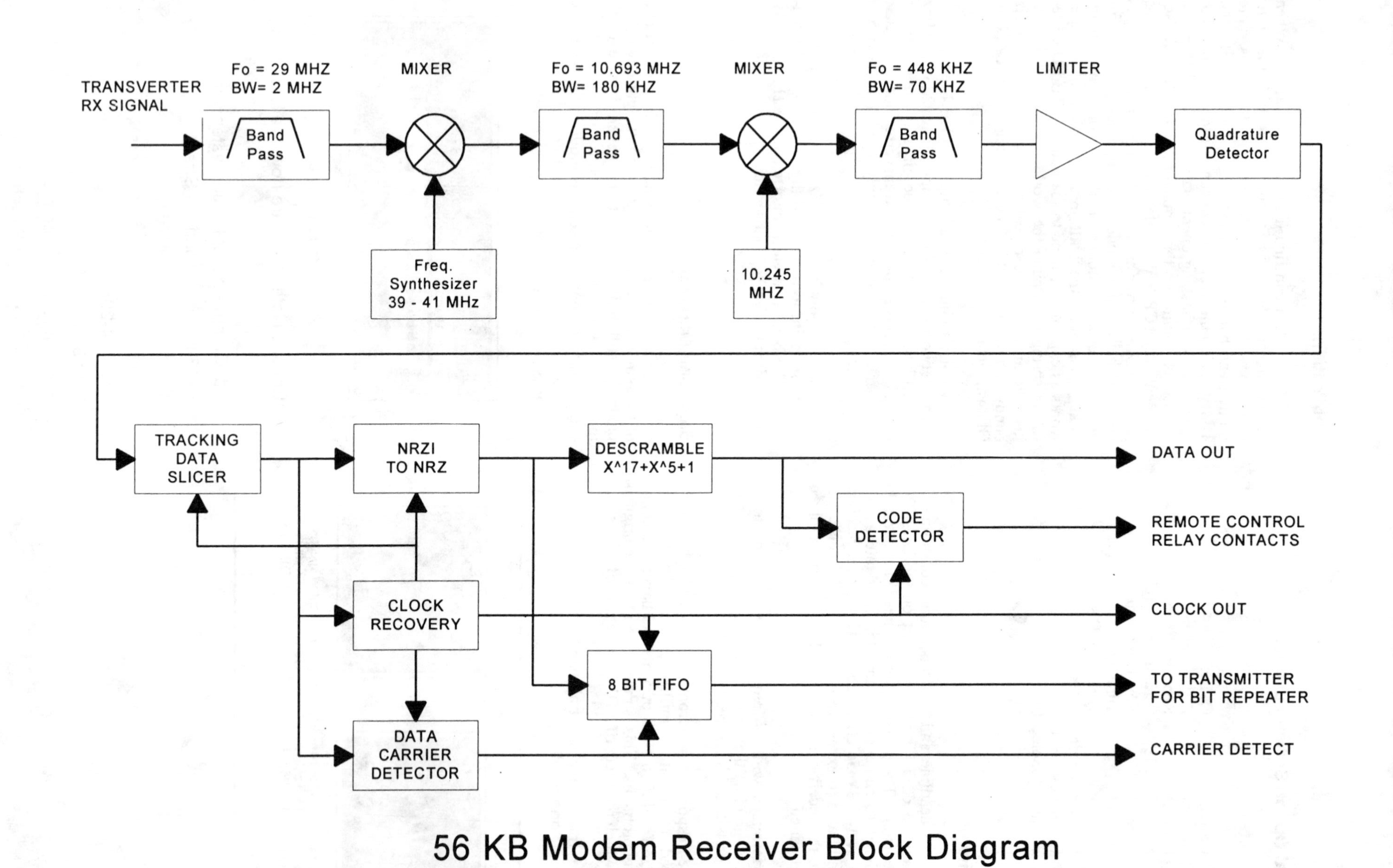

56 KB Modem Receiver Block Diagram

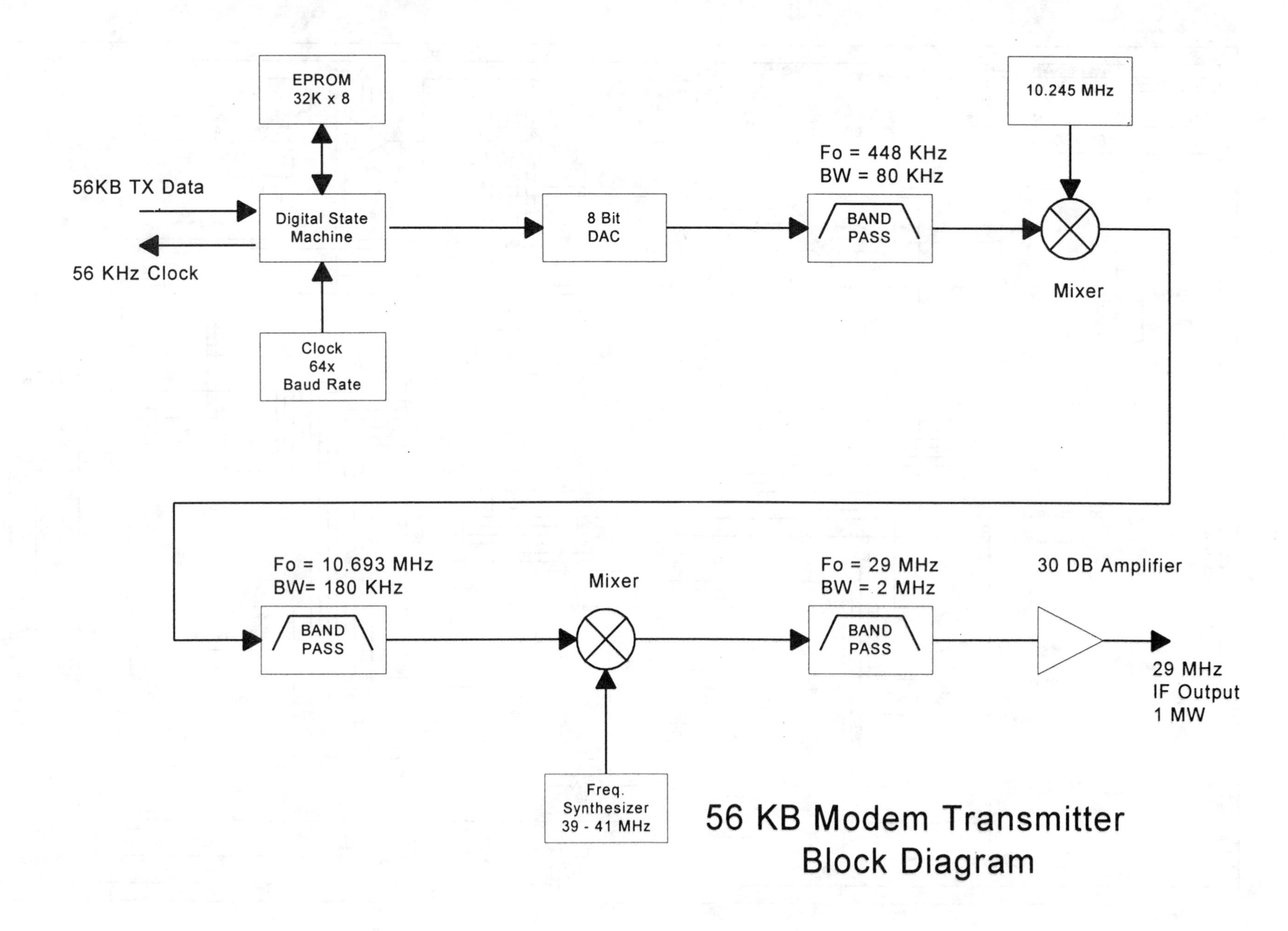

EPROM
32K x 8
10.245 MHz
56KB TX Data
56 KHz Clock
Digital State
Machine
8 Bit
DAC
Fo = 448 KHz
BW = 80 KHz
BAND
PASS
Mixer
Clock
64x
Baud Rate
Fo = 10.693 MHz
BW= 180 KHz
BAND
PASS
Mixer
Fo = 29 MHz
BW = 2 MHz
BAND
PASS
30 DB Amplifier
29 MHz
IF Output
1 MW
Freq.
Synthesizer
39 - 41 MHz
56 KB Modem Transmitter
Block Diagram

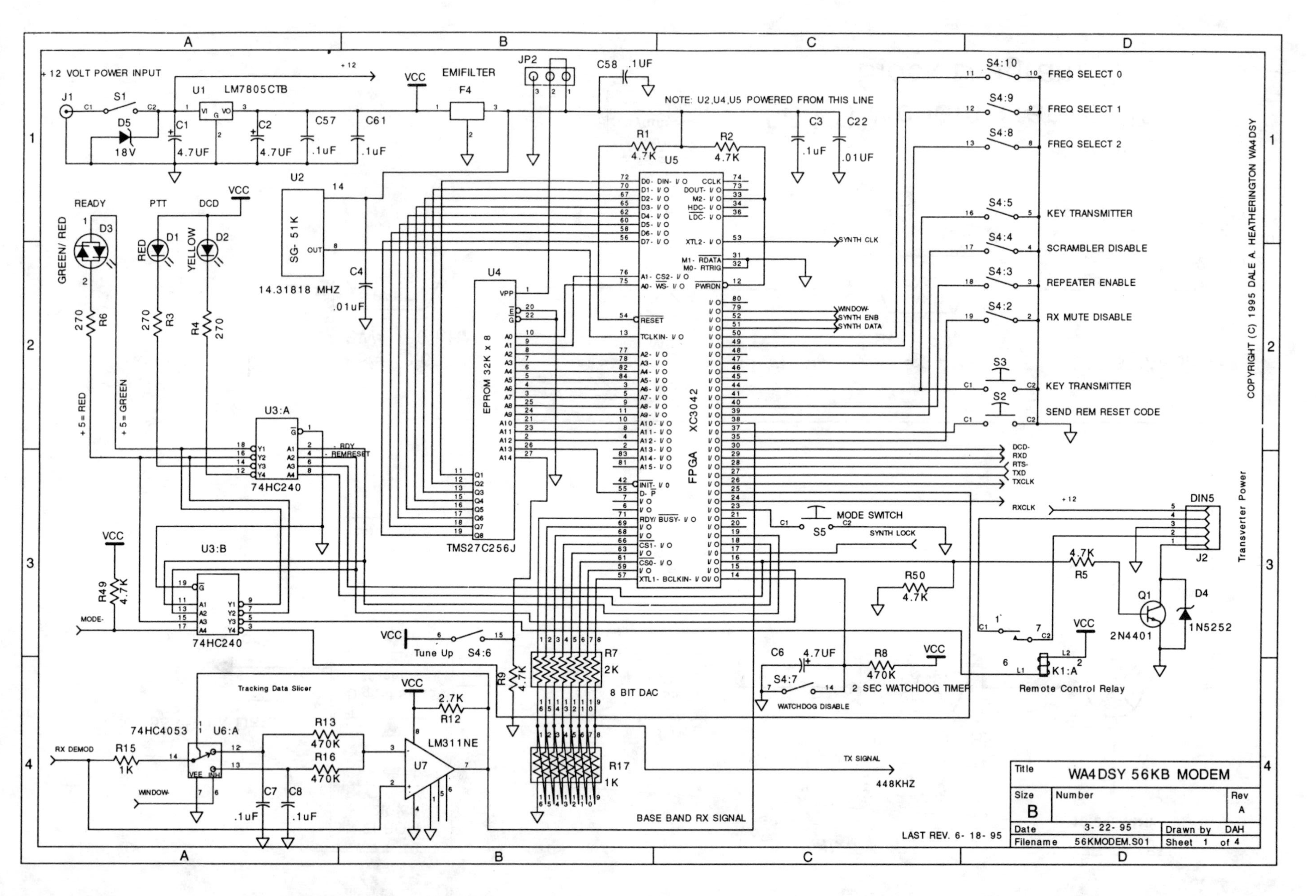

+12 VOLT POWER INPUT
J1
S1
D5 18V
U1 LM7805CTB
C1 4.7UF
C2 4.7UF
C57 .1uF
C61 .1uF
+12
VCC
EMIFILTER F4
JP2
C58 .1UF
NOTE: U2,U4,U5 POWERED FROM THIS LINE
C3 .1uF
C22 .01UF
R1 4.7K
R2 4.7K
U5
FPGA XC3042
U2
SG-51K
14.31818 MHZ
C4 .01uF
U4
EPROM 32K x 8
TMS27C256J
READY
PTT
DCD
D3 GREEN/RED
D1 RED
D2 YELLOW
R6 270
R3 270
R4 270
+5 = RED
+5 = GREEN
U3:A
74HC240
U3:B
74HC240
RDY
REMRESET
VCC
R49 4.7K
MODE-
SYNTH CLK
WINDOW-
SYNTH ENB
SYNTH DATA
S4:10 FREQ SELECT 0
S4:9 FREQ SELECT 1
S4:8 FREQ SELECT 2
S4:5 KEY TRANSMITTER
S4:4 SCRAMBLER DISABLE
S4:3 REPEATER ENABLE
S4:2 RX MUTE DISABLE
S3 KEY TRANSMITTER
S2 SEND REM RESET CODE
COPYRIGHT (C) 1995 DALE A. HEATHERINGTON WA4DSY
DCD-
RXD
RTS-
TXD
TXCLK
RXCLK
MODE SWITCH
S5
SYNTH LOCK
DIN5
J2
Transverter Power
R5 4.7K
R50 4.7K
Q1 2N4401
D4 1N5252
VCC
K1:A
Remote Control Relay
Tune Up S4:6
R9 4.7K
R7 2K
8 BIT DAC
C6 4.7UF
S4:7
WATCHDOG DISABLE
R8 470K
2 SEC WATCHDOG TIMER
Tracking Data Slicer
74HC4053
U6:A
RX DEMOD
R15 1K
WINDOW-
R13 470K
R16 470K
C7 .1uF
C8 .1uF
R12 2.7K
LM311NE
U7
R17 1K
BASE BAND RX SIGNAL
TX SIGNAL
448KHZ
LAST REV. 6-18-95
Title WA4DSY 56KB MODEM
Size B
Number
Rev A
Date 3-22-95
Drawn by DAH
Filename 56KMODEM.S01
Sheet 1 of 4

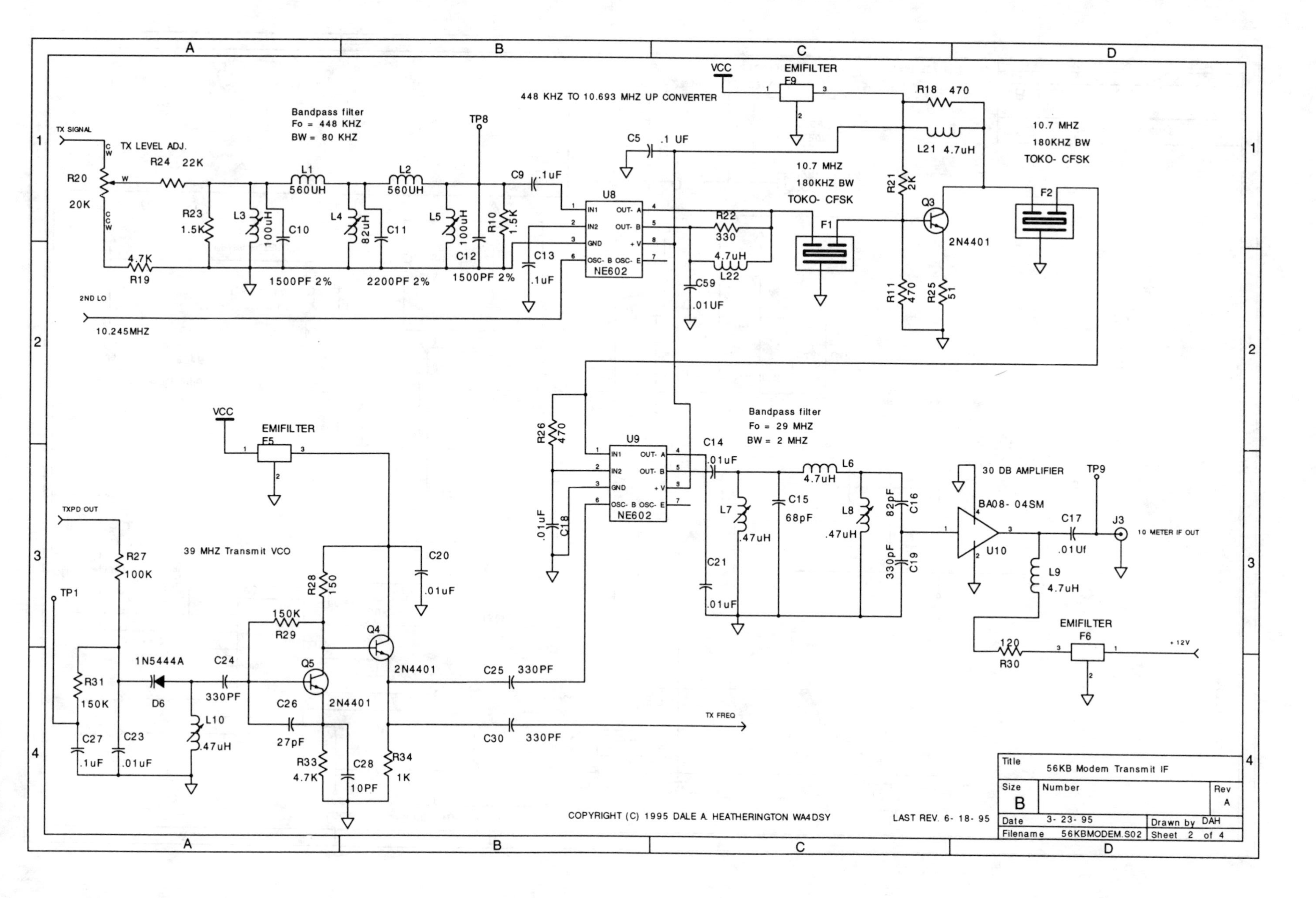
448 KHZ TO 10.693 MHZ UP CONVERTER
Bandpass filter
Fo = 448 KHZ
BW = 80 KHZ
TX SIGNAL
TX LEVEL ADJ.
R24 22K
R20
20K
R23
1.5K
R19
4.7K
L1
560UH
L2
560UH
L3
100uH
L4
82uH
L5
100uH
C10
C11
C12
1500PF 2%
2200PF 2%
1500PF 2%
TP8
R10
1.5K
C9 .1uF
C13
.1uF
2ND LO
10.245MHZ
C5 .1 UF
VCC
EMIFILTER
F9
U8
NE602
R22
330
4.7uH
L22
C59
.01UF
10.7 MHZ
180KHZ BW
TOKO- CFSK
F1
R18 470
L21 4.7uH
R21
2K
Q3
2N4401
R11
470
R25
51
10.7 MHZ
180KHZ BW
TOKO- CFSK
F2
R26
470
U9
NE602
.01uF
C18
Bandpass filter
Fo = 29 MHZ
BW = 2 MHZ
C14
.01uF
L6
4.7uH
L7
.47uH
C15
68pF
L8
.47uH
82pF
C16
330pF
C19
C21
.01uF
30 DB AMPLIFIER
BA08- 04SM
U10
TP9
C17
.01Uf
J3
10 METER IF OUT
L9
4.7uH
EMIFILTER
F6
120
R30
+12V
VCC
EMIFILTER
F5
TXPD OUT
R27
100K
TP1
39 MHZ Transmit VCO
R28
150
C20
.01uF
150K
R29
Q4
Q5
2N4401
2N4401
1N5444A
D6
C24
330PF
R31
150K
L10
.47uH
C27
.1uF
C23
.01uF
C26
27pF
R33
4.7K
C28
10PF
R34
1K
C25 330PF
C30 330PF
TX FREQ
COPYRIGHT (C) 1995 DALE A. HEATHERINGTON WA4DSY
LAST REV. 6- 18- 95
Title 56KB Modem Transmit IF
Size B
Number
Rev A
Date 3- 23- 95
Drawn by DAH
Filename 56KBMODEM.S02
Sheet 2 of 4

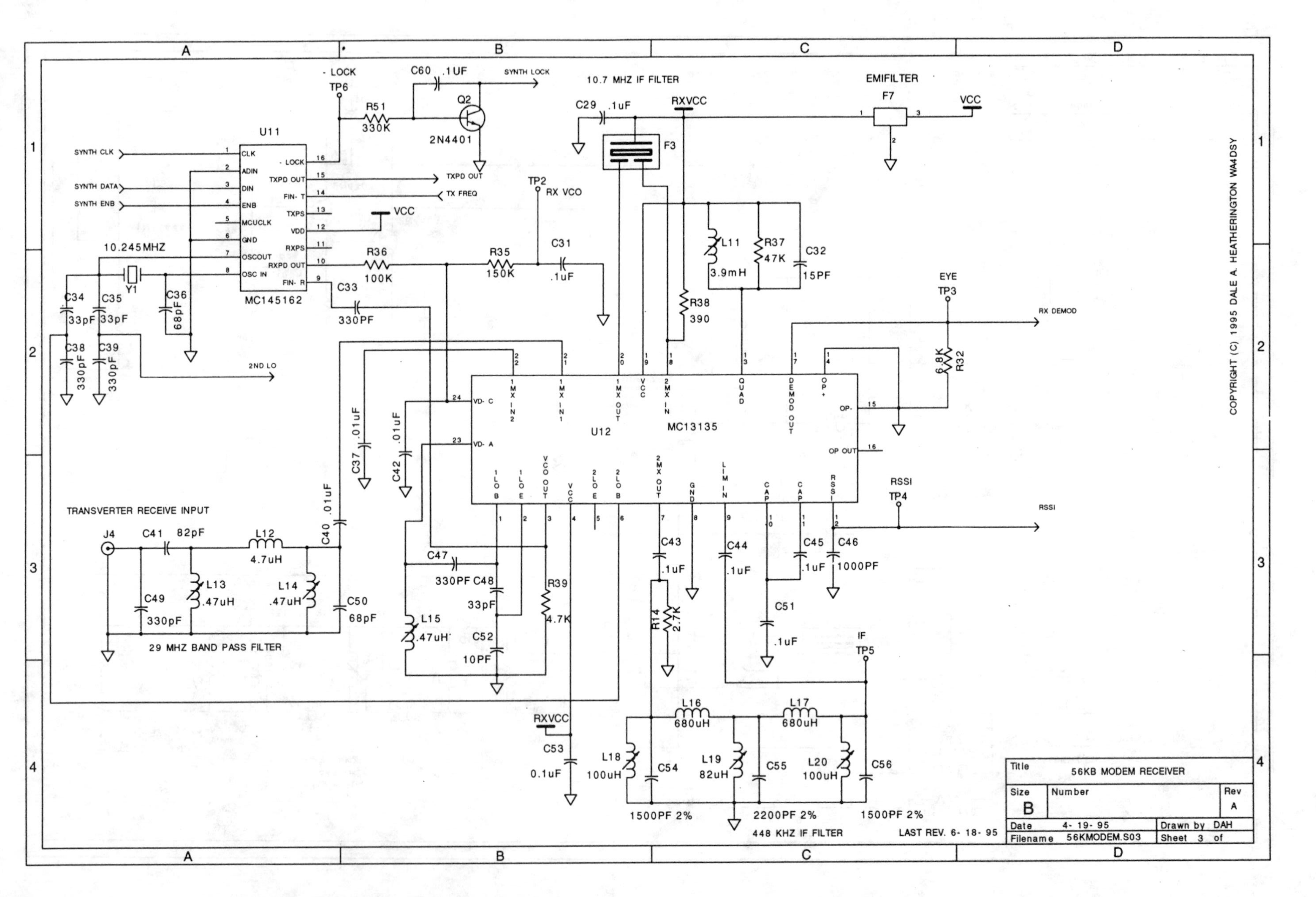

- LOCK
TP6
R51
330K
C60 .1UF
SYNTH LOCK
Q2
2N4401
10.7 MHZ IF FILTER
C29 .1uF
RXVCC
F3
EMIFILTER
F7
VCC
U11
SYNTH CLK
SYNTH DATA
SYNTH ENB
CLK
ADIN
DIN
ENB
MCUCLK
GND
OSCOUT
OSC IN
- LOCK
TXPD OUT
FIN- T
TXPS
VDD
RXPS
RXPD OUT
FIN- R
MC145162
TXPD OUT
TX FREQ
VCC
TP2
RX VCO
10.245MHZ
Y1
C34 33pF
C35 33pF
C36 68pF
C38 330pF
C39 330pF
2ND LO
R36 100K
R35 150K
C31 .1uF
C33 330PF
L11 3.9mH
R37 47K
C32 15PF
R38 390
EYE
TP3
RX DEMOD
R32 6.8K
U12
MC13135
VD- C
VD- A
1 MX IN 2
1 MX IN 1
1 MX OUT
VCC
2 MX IN
QUAD
DEMOD OUT
OP +
OP-
OP OUT
1 LO B
1 LO E
VCO OUT
VCC
2 LO E
2 LO B
2 MX OUT
GND
LIM IN
CAP
CAP
RSSI
RSSI
TP4
RSSI
C37 .01uF
C42 .01uF
TRANSVERTER RECEIVE INPUT
J4
C41 82pF
L12 4.7uH
C40 .01uF
C49 330pF
L13 .47uH
L14 .47uH
C50 68pF
29 MHZ BAND PASS FILTER
C47 330PF
C48 33pF
R39 4.7K
L15 .47uH
C52 10PF
C43 .1uF
R14 2.7K
C44 .1uF
C45 .1uF
C46 1000PF
C51 .1uF
IF
TP5
RXVCC
C53 0.1uF
L16 680uH
L17 680uH
L18 100uH
C54
L19 82uH
C55
L20 100uH
C56
1500PF 2%
2200PF 2%
1500PF 2%
448 KHZ IF FILTER
LAST REV. 6- 18- 95
COPYRIGHT (C) 1995 DALE A. HEATHERINGTON WA4DSY
Title 56KB MODEM RECEIVER
Size B
Number
Rev A
Date 4- 19- 95
Drawn by DAH
Filename 56KMODEM.S03
Sheet 3 of

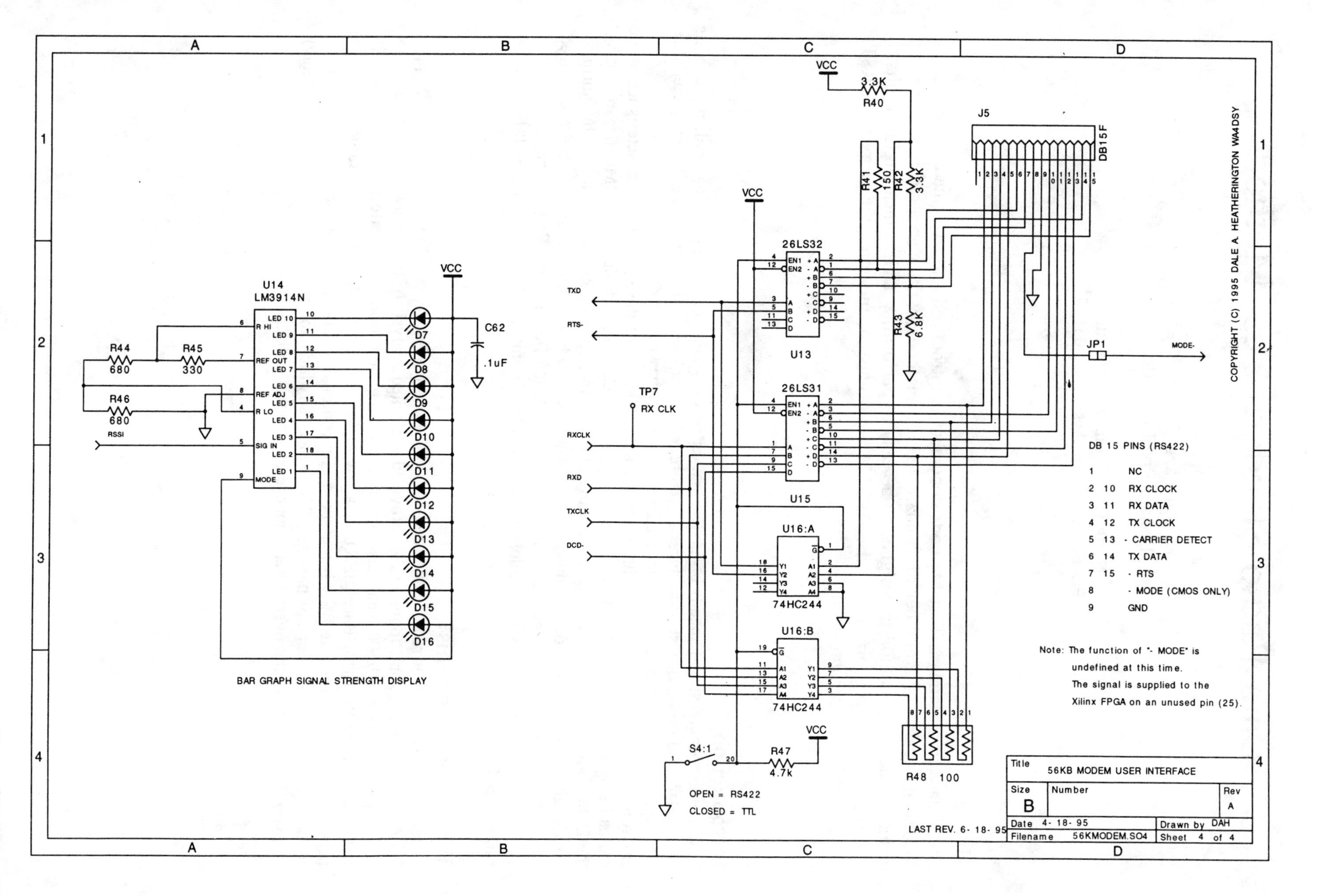
COPYRIGHT (C) 1995 DALE A. HEATHERINGTON WA4DSY
DB 15 PINS (RS422)
1 NC
2 10 RX CLOCK
3 11 RX DATA
4 12 TX CLOCK
5 13 - CARRIER DETECT
6 14 TX DATA
7 15 - RTS
8 - MODE (CMOS ONLY)
9 GND
Note: The function of "- MODE" is undefined at this time.
The signal is supplied to the Xilinx FPGA on an unused pin (25).
Title 56KB MODEM USER INTERFACE
Size B
Number
Rev A
Date 4- 18- 95
Drawn by DAH
Filename 56KMODEM.SO4
Sheet 4 of 4
LAST REV. 6- 18- 95
J5
DB15F
JP1
MODE-
VCC
R40 3.3K
R41 150
R42 3.3K
R43 6.8K
R48 100
26LS32
U13
26LS31
U15
U16:A
U16:B
74HC244
R47 4.7k
S4:1
OPEN = RS422
CLOSED = TTL
TP7 RX CLK
TXD
RTS-
RXCLK
RXD
TXCLK
DCD-
U14
LM3914N
C62 .1uF
D7 D8 D9 D10 D11 D12 D13 D14 D15 D16
R44 680
R45 330
R46 680
RSSI
BAR GRAPH SIGNAL STRENGTH DISPLAY

Data Radio Standard Test Methods

Burton Lang, VE2BMQ
Donald Rotolo, N2IRZ

North East Digital Association
PO Box 563, Manchester NH 03105

Abstract
The Data Radio Standard Test Methods document is introduced and explained. The document consists of a number of standardized test methods, written in a clear, step-by-step format. Each test method is designed to be easy to perform, yet yield meaningful results. The rationale and organization of the DRSTM are discussed, including the proposed DRSTM database of measurement data.

Introduction
With the ever-increasing demands of modern packet networks upon the radios being used, there exists no standardized group of test procedures for these radios, or even a list of which radio characteristics are important, and why. If we are to maximize the efficiency of the shared radio networking environment we use, then we must have a better understanding of the significant characteristics of the radios being used. In an effort to resolve this apparent lack of information, VE2BMQ began writing the Data Radio Standard Test Methods in August 1994. In October 1994, the Board of the North East Digital Association endorsed this effort as beneficial to advancing the radio art, and encouraged other organizations to do so as well. In November 1994, the Board of the Radio Amateur Telecommunications Society also endorsed the DRSTM. In this paper, these efforts will be summarized.

Why do we need this?
Recently, a test report of radios advertised as "9600 Baud ready" appeared in QST [May 1995, pp 24-29]. The author, Jon Bloom KE3Z, provided a detailed analysis of the bandwidth considerations for proper operation at 9600 baud, also briefly mentioning deviation and its effects. He then provided test results of the Bit Error Rate (BER) for a few radios, a critical yardstick of data radio performance. However, the BER is measured in a continuous data stream, without the normally found pauses between groups of packets. These pauses, which are set by TXDelay (transmit delay) can significantly affect the overall performance of a packet radio link.

As an example, compare two commonly used 70 cm radios: The TEKK T-Net Mini (Model KS-900L) and the ICOM IC-449A. The TEKK has a RX/TX turnaround time of less than 12 milliseconds, while the ICOM has been measured at over 410 milliseconds. At 9600 baud, this translates to a loss of data capacity on the order of 80%. From the table below, we can see that a 2400 baud link with a TXDelay of 40 mS has a higher throughput capacity than a 56 kb link with a TXDelay of 500 mS!

Baud	Byte	Throughput (bps) given a TXDelay of:				
Rate	Time	0mS	40mS	250mS	350mS	500mS
1200	6.67mS	127	121	99	91	81
2400	3.33mS	254	233	163	143	121
4800	1.67mS	506	431	241	199	158
9600	.833	1015	750	316	248	187
56k	.104	8131	2124	435	316	223

[Reprinted from NEDA 1994 Annual, pg.81. Assumptions: 230 Byte data per 256 byte transmission, 16 byte acknowledgement, both transmitters have same TXD]

The point is that one parameter that was not even measured in the test has a huge influence upon the performance of a data radio in a real packet network. If such a serious omission can be found in a technically competent journal such as QST, it can be said that most amateurs have an incomplete understanding of the parameters affecting data radio performance.

Organization

The Data Radio Standard Test Methods (DRSTM) is organized much like the published procedures of large standards organizations, such as the ASTM, IEEE or SAE. Each test method first defines the parameter to be measured, then explains its importance to data transmission. A detailed, step-by-step test procedure is then provided, along with set-up diagrams and a standardized form for recording and interpreting the test results. This ensures that all test results, whatever their source, are as reliable as possible. Efforts are taken to use commonly available test equipment wherever possible. This allows as many people as possible to perform the tests. The tests themselves are kept as simple as possible, while still yielding meaningful results, which allows a wide range of radio equipment to be tested. While we believe that many radio users would not actually perform measurements, just knowing how a particular parameter is measured can offer insight to its effects.

In addition, the DRSTM manual contains a detailed Glossary, a thorough explanation of the database structure, and performance requirements for all test equipment.

The scope and significance of each test that has been written so far is summarized below:

DRSTM-01 Transmitter Power-on Time Delay

Scope: This test procedure is intended to measure the time from the start of the transmitter keying (Push-To-Talk or PTT) line becoming active until the RF output has risen to 90% of its final value.

Significance: The time delay that a radio transmitter's power output has when it is keyed can range from microseconds to hundreds of milliseconds. A transmitter that is keyed but not putting out RF power can create a serious 'hidden transmitter' problem. In addition, in the case of a synthesized transmitter, a significant difference between the Transmitter Power-on Time Delay and the Transmitter Power On-frequency Time Delay (see DRSTM-02) would indicate the possibility of serious interference to users on other nearby frequencies.

DRSTM-02 Transmitter Power On-frequency Time Delay, Frequency Stability Time Delay, and Modulation Stability Time Delay

Scope: This test procedure will measure three Data Radio characteristics:
1. Time from PTT becoming active to the RF output appearing within the designated passband of a test receiver.
2. Time delay until the transmitter output frequency has stabilized to within 5% of the channel bandwidth relative to its final stable frequency, or until the PLL loop tone or any other extraneous signal has decayed to less than 10% of the normal system modulation level, whichever is longer.
3. The time delay until the transmitter's modulation envelope has reached 90% of its final stable value.

Significance: The time delay until a transmitter's output frequency appears within its intended channel can range between microseconds and hundreds of milliseconds. Any difference between the Power Output Time Delay (see DRSTM-01) and the Power Output On-frequency Time Delay would indicate the possibility of serious interference to users on nearby frequencies. Any substantial delay in the appearance of the transmitter output signal in the receivers of other users on the same channel can result in a serious 'hidden transmitter' problem.

The additional delay of waiting for the frequency to settle down, or signals that have a significant amount of non-data modulation (such as the damped oscillation of a PLL oscillator feedback loop immediately following lockup) can have a serious effect on decoding the data signal. This could require a much longer TXDelay setting for usable operation.

Similarly, the time delay for the modulation envelope to settle down to its final stable value can also affect proper data recovery at the beginning of a data transmission. This rare condition has been observed in certain radios such as the Motorola MOCOM 35, where a large RC time constant in the reactance modulator retards the modulation envelope by several hundred milliseconds.

DRSTM-03 Transmitter Short-term Frequency Stability

Scope: This test procedure is intended to measure the transmitter and receiver frequency changes that occur when the radio is exposed to extreme temperature and supply voltage conditions.

Significance: Excessive changes of either the transmitter or receiver channel frequency can cause failure or degradation of packet radio links, if the change forces the signal beyond the passband of the received at either end. This information is important to designers of packet network facilities where equipment may be housed in unheated enclosures, as well as builders of emergency response networks that must not fail under extreme conditions.

It must be recognized that the quartz crystal used as the primary channel element or as a reference for a PLL is a major factor in frequency stability. Amateur radio users often use crystals from suppliers that may not follow the manufacturer's specifications. testing radios using such crystals can lead to more appropriate selection of crystals for particular radios.

DRSTM-04 Receiver Detector/Frequency/Squelch Recovery Time Delay

Scope: This test procedure measures:
1. The time between when PTT is released until a signal is received at the detector output
2. The time between when PTT is released until the receiver center frequency is within 5% of its design bandwidth relative to its final stable receive frequency, or until the PLL loop tone or other extraneous signal has decayed to within 10% of the normal demodulated data signal amplitude, whichever is longer
3. The time between when PTT is released until the receiver squelch circuit (and all preceding circuits) begin to pass a demodulated signal.

Significance: A useful data signal cannot be recovered from a received signal until the detector and all RF/IF/LO circuits ahead of it have recovered following a period of transmission.

Radios which take a long time for their frequency to stabilize, or that have considerable 'non-data' signals (such as the damped oscillation of a PLL oscillator feedback loop immediately following lockup) can have a serious effect on decoding the data signal. This could require a much longer TXDelay setting for usable operation.

Radio squelch circuits which take a long time to open following a transmit period would present decoding difficulties when the TNC or modem is connected after the squelch circuit (such as at the speaker jack). This would also require a longer setting for TXDelay.

DRSTM-05 Receiver Squelch Turn-on Time Delay

Scope: This test procedure measures the time between the start of a test signal until the squelch circuit in the radio opens and passes demodulated audio.

Significance: When using a TNC or modem connected after the squelch circuit, the time it takes for the squelch circuit to open and pass audio affects operation. If there is a significant time delay in opening the squelch, the TNC could decide to transmit before it realized the channel was occupied. This would cause a 'hidden transmitter' like problem.

DRSTM-08 Receiver Output Level, Impedance, Demodulated Frequency Slope (De-emphasis) and Demodulation 6dB Cutoff Bandwidth.

Scope: This test procedure measures:
1. The voltage level of the demodulated data signal output, per unit modulation level.
2. The impedance at the point where the demodulated audio output signal is connected.
3. The amount of audio output signal variation with frequency (De-emphasis slope) measured where the output is connected, expressed as dB per octave.
4. The audio frequency at which the demodulated signal drops to one-half the voltage on both the higher and lower sides of the normal operating frequency range.

Significance: The level and impedance of the demodulated audio signal from a data radio receiver is useful information when designing data systems and interfacing TNCs and modems. The de-emphasis response of a receiver affects TNC or modem operation. Most TNCs and modems require a flat frequency response while most radios offer some pre- and de-emphasis to improve voice quality. For optimum performance, these responses should be matched. The high side frequency response helps determine the maximum usable bit rate, while the low side frequency response is important with certain direct FM modulation systems (such as the G3RUH modem).

Editors Note:
the latest information for this project may be found on the Web at:
http://www.rocler.qc.ca/burt/drstm.html.

DRSTM-09 Transmitter Modulation Drive Requirements, Input Impedance, Impedance Response Slope, Pre-emphasis Slope, Maximum Deviation and Modulation Frequency Capability.

Scope: This test procedure measures:
1. The voltage level of the modulation signal per unit of transmitter deviation.
2. The input impedance of the modulation circuit.
3. The response characteristics of the modulator input impedance, as it varies with audio frequency.
4. The variation in the deviation when the modulating frequency is changed (pre-emphasis).
5. The maximum deviation capability without distortion.
6. The highest modulating frequency which does not reduce the first modulation sideband by more than 20dB.

Significance: The level and impedance of the modulator is useful when designing data systems and interfacing TNCs and modems. The modulator frequency response characteristic should be matched with the receiver response to obtain an overall 'flat' system response. A non-flat system response will distort data signals. The impedance variation of the modulator with frequency can also affect the modulator's frequency response characteristic. The maximum modulation (deviation) capability without distortion is useful when designing data systems and interfacing TNCs and modems. The maximum modulating frequency is useful in determining the maximum capabilities of a higher-speed data link, as well as in avoiding the radiation of wide sidebands produced by computer clock noise, etc. FM transmitters with direct connections to their modulators have been found radiating 10 MHz wide sidebands, caused by leakage of the TNCs 5 MHz CPU clock.

Database

An integral part of the DRSTM concept is a database containing the measurement results obtained by the DRSTM users. Each DRSTM enables the user to make the same measurements, consistently, and provides a form on which to record all measurements. It is well known that radios have differing characteristics for many parameters, even radios of the exact same type and of consecutive serial numbers. It is anticipated that, by providing an international clearinghouse and database of all measurements, that the amateur community would be better served. this may also provide an incentive for radio manufacturers to either publish the data-relevant parameters, or at least design radios with data transmission in mind. While no firm plans have been developed, it is expected that the database would be available on-line in some fashion, with free access for all.

Conclusion

The establishment of standards for various radio characteristics having significance in data transmission will eliminate much of the confusion and misinformation in the area today. These standard test methods, used to measure the performance of data radios, could be used by anyone having reasonable experience with common electronic test equipment. The international database would disseminate this collected data and, as manufacturers noticed that certain radios were unsuitable for data use, convince radio designers to modify their designs to acommodate data transmission. In this manner, the authors hope to improve the radio art.

It is the author's hope that other knowledgeable persons would step forward and offer their expertise in either writing standard methods, suggesting new tests, evaluating existing tests, or performing tests and disseminating their findings. At this time, these standard tests should be considered tentative. If you can help in any way, please contact the authors.

javAPRS

Implementation of the APRS Protocols in Java

Steve Dimse KO4HD

sdimse@bridge.net http://www.bridge.net/~sdimse

This paper describes my implementation of the Automatic Position Reporting System (APRS) protocols in the computer language Java. APRS is one of the most innovative uses of ham radio in recent years. javAPRS extends the usefulness of APRS to the internet.

Java Basics

Java was designed by Sun Microsystems as a language to promote the use of distributed computing resources over the internet. Based on the C language, with an object programming philosophy drawn from Smalltalk and Lisp, platform independence, and built-in security, it is a language uniquely suited to network programming.

The principle drawback to the use of Java at present is slow execution speed. Unlike traditional programs, which are compiled into machine specific object code before distribution, Java is compiled down to *byte code* which is then interpreted on the local machine, often executing at 10-25% of the speed of native code. There is reason to be hopeful, however. Several companies are working on 'just-in-time' compilers, which convert byte code to native code on the local machine just prior to execution. Also, the horsepower of today's computers makes even the interpreted version run at an acceptable speed in most applications.

The leading use of Java at present is to write programs, called *applets*, which are run within a World Wide Web (WWW) browser, such as Netscape. javAPRS can work in this fashion, or as a stand alone program. Security restrictions limit some of the interesting possibilities for javAPRS as an applet, such as using separate servers for map and APRS data, or connecting to multiple other computers to plot data from multiple LANs.

javAPRS Design

In the basic design of javAPRS, I have tried to create a system which can be used now by people without programming knowledge to add APRS data to their web pages. In addition, using the object oriented programming (OOP) features of Java, the system is designed to be easily extended by other Java programmers. This sort of extension does not require access to source code of javAPRS. The interfaces used by other programs will be posted in my web pages as they are finalized. Source code will not be freely available, but I will consider requests for the source code on a specific basis. The remainder of this paper discusses the use of javAPRS in web page creation. Programmers interested in extending javAPRS should contact me directly for more info.

javAPRS Applet Parameters

The basic syntax to call an applet in hypertext markup language (HTML) is:

```
<APPLET codebase = "javAPRS/" CODE="javAPRS.class" WIDTH=400 HEIGHT =300>
```

This executes the applet "javAPRS.class" in the directory "javAPRS" (relative to the HTML file the call is found in). It allocates a space of the size indicated in the browser window. Other HTML commands, such as <CENTER> work just as they would for other graphic elements such as images. The lines after APPLET tag consist of a series of parameters which are passed to the applet to control its behavior. Each parameter has a default value, usually the most commonly chosen option, and if the default parameter is desired, it need not be declared.

Map Parameters

At this time, javAPRS understands two kinds of maps. One or the other map must be used, or the applet will not run. The map files are stored in a subfolder "/maps" relative to the "codebase" named in the applet call.

```
<PARAM name = "dosMap" value = "anymap.map">
<PARAM name = "gifMap" value = "anymap.map">
```

Maps can be automatically or manually scaled.

```
<PARAM name = "autoScale" value  = "true">  (Default  true)
```

This will cause the map to be scaled to fit the window the applet is presently running in. If autoscaling is not used, then the following three parameters may be used to set the magnification and offset of the map. For now, the only way to figure out the value of these parameters is trial and error.

```
<PARAM name = "scale" value   = "2.0">  (Default  1.0)
<PARAM name = "offsetX" value  = "100">  (Default  0)
<PARAM name = "offsetY" value  = "100">  (Default  0)
```

Two options work only with dosMaps, namely:

```
<PARAM name = "showMapLabels" value = "true">   (default true)
<PARAM name = "showAllMapLabels" value = "true">   (default false)
```

which will show either all map labels or those designated at the present scale.

Data Parameters

Data to be displayed by javAPRS is one of three types, either NMEA (only RMC and GGA are recognized at present), TNC data (raw data from a TNC which is the MacAPRS log file format), and HST files produced by dosAPRS. Any or all of the three types of data may be displayed, but only one file of each type can be used.

```
<PARAM name = "NMEAfile" value = "NMEA.data">
<PARAM name = "TNCfile" value = "ko4hd.data">
<PARAM name = "HSTfile" value = "marathon.data">
```

The way the data is displayed is affected by several parameters:

```
<PARAM name = "displayVectors" value = "true">   (default  true)
```

This will draw vectors for course and speed info if present in a position report.

```
<PARAM name = "showCallsigns" value = "true">   (default  true)
```

This prints a callsign next to each position report.

```
<PARAM name = "stationList" value = "true">
```

javAPRS can keep a station list for the stations that are 'heard' in the data stream. At the end of a file readin, the contents are dumped to the java console (select the option "Open Java Console" in Netscape. It will also speed up the redrawing once the data have been read, only the last position of each station will be plotted.

```
<PARAM name = "showNewStations" value = "false">
```

If stationList is true, then this option will show the name of each new station as it is heard. If the home station has been specified, the bearing and distance to the station will also be displayed.

There are a number of other, less important parameters that are available to fine tune the display to suit the user. Please refer to my web pages for more details.

javAPRS Sample Applications

The javAPRS classes that exist now allow a person without programming skills to create a World Wide Web page containing an APRS map, and display a file containing positions of various APRS stations, objects, and track plots. More complete instructions may be found on my web site. Here are three examples, with the HTML code used to call the applet, and the URL's to reference them.

HST file replay

(http://www.bridge.net/~sdimse/marathon.html)

```
<APPLET codebase = "javAPRS/" CODE="javAPRS.class" WIDTH=500 HEIGHT =350>
<PARAM name = "dosMap" value = "washdc.map">
<PARAM name = "HSTfile" value = "marathon.hst">
<PARAM name = "sleep" value = "50">
<PARAM name = "stationlist" value = "false">
<PARAM name = "showStationNames" value = "false">
<PARAM name = "copyrightTop" value = "false">
<PARAM name = "scale" value = "1.6">
<PARAM name = "offsetx" value = "620">
<PARAM name = "offsety" value = "310">
</APPLET>
```

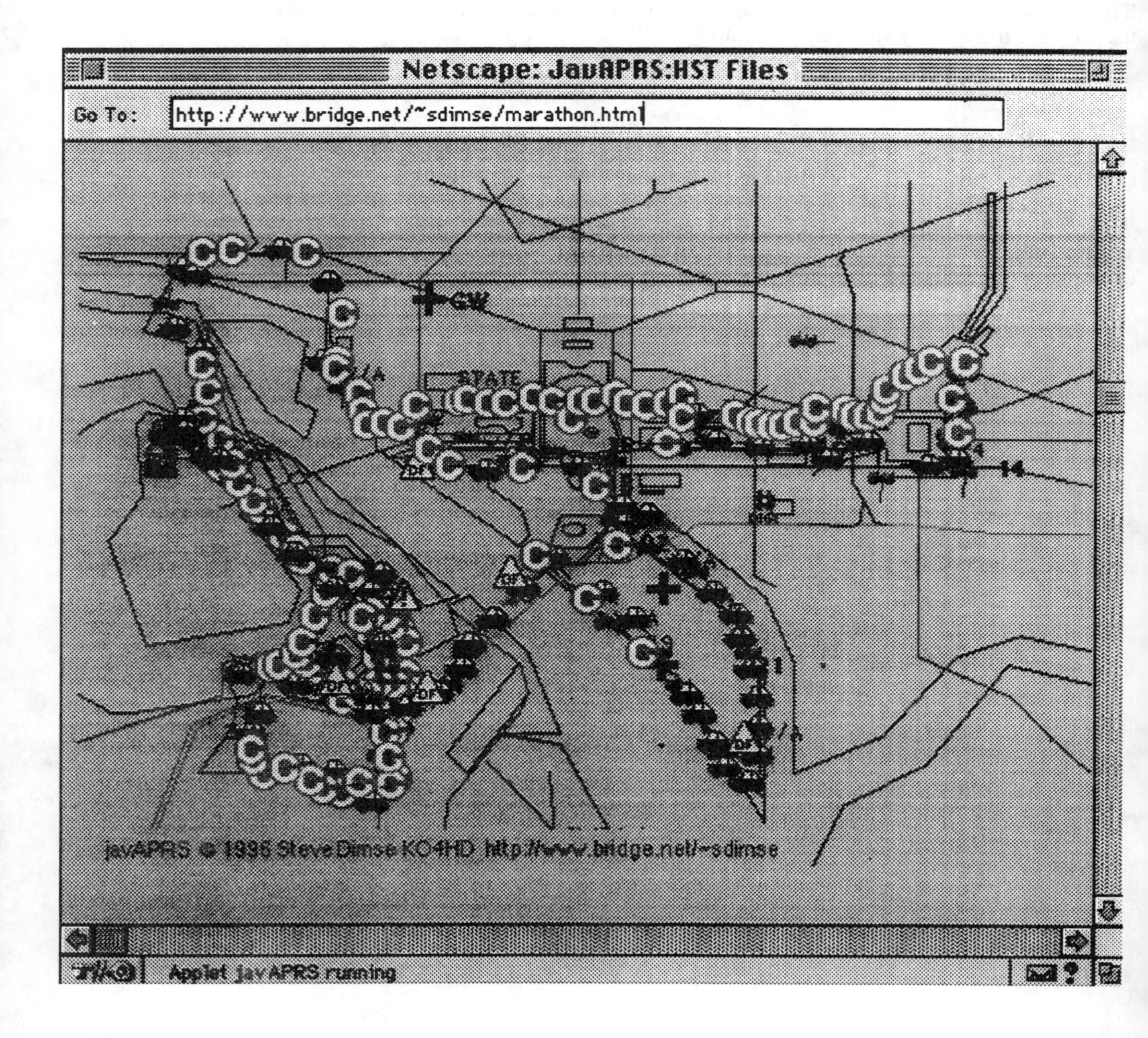

GIF files for maps

(http://www.bridge.net/~sdimse/gifmap.html)

```
<APPLET codebase = "javAPRS/" CODE="javAPRS.class" WIDTH=500 HEIGHT =350>
<PARAM name = "gifMap" value = "cudjoe.gif">
<PARAM name = "gifMapLeft" value = "81.558">
<PARAM name = "gifMapTop" value = "24.7">
<PARAM name = "gifMapPPDh" value = "3800">
<PARAM name = "gifMapPPDv" value = "3900">
<PARAM name = "sleep" value = "700">
<PARAM name = "LLfile" value = "boattrip.ll">
<PARAM name = "stationlist" value = "false">
<PARAM name = "copyrightTop" value = "false">
</APPLET>
```

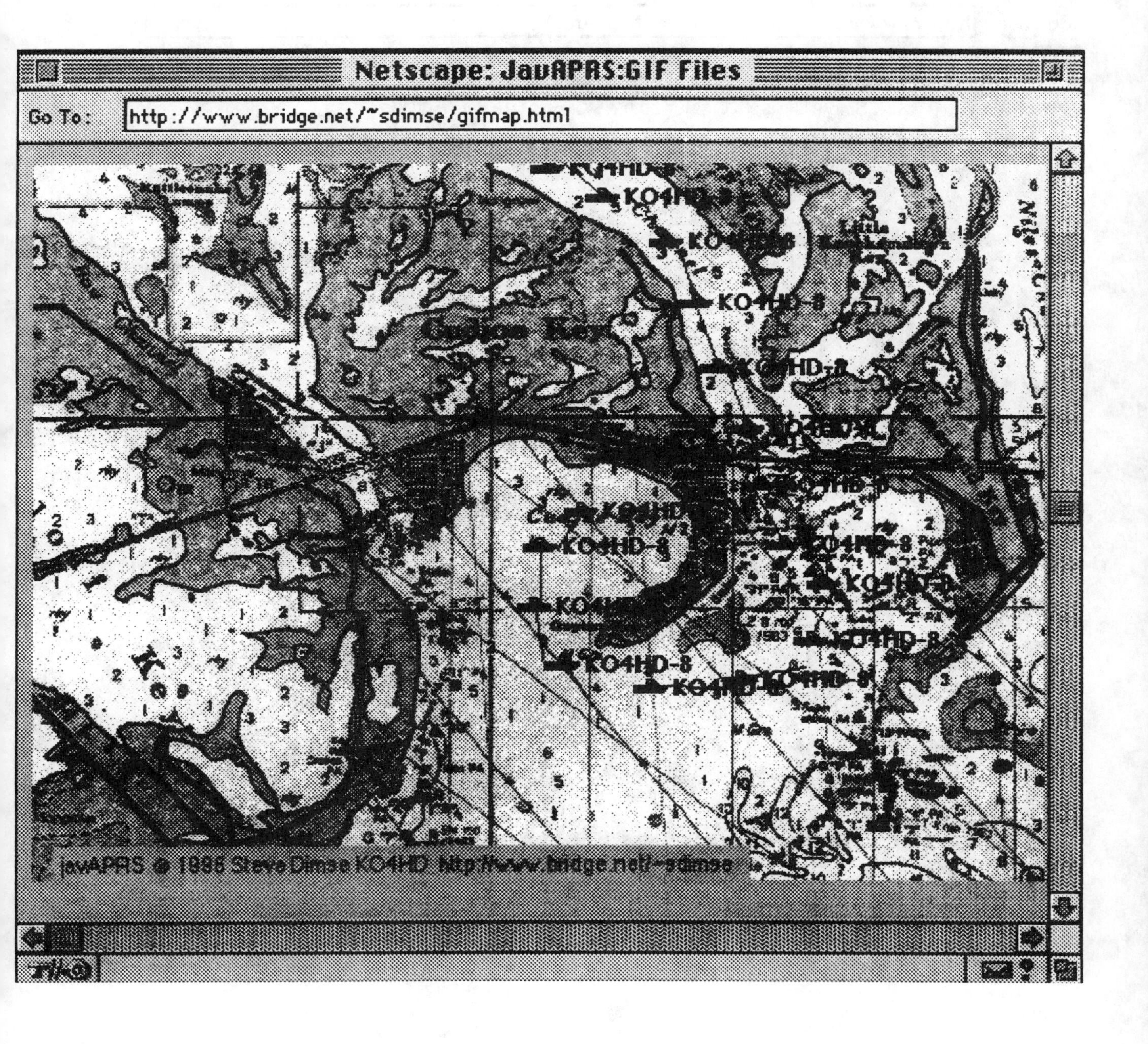

Display of TNC data capture

(http://www.bridge.net/~sdimse/trip.html)

```
<APPLET codebase = "javAPRS/" CODE="javAPRS.class" WIDTH=500 HEIGHT =350>
<PARAM name = "dosMap" value = "usa.map">
<PARAM name = "TNCfile" value = "tnc.data">
<PARAM name = "drawvectors" value = "true">
<PARAM name = "homeID" value = "KO4HD-9">
</APPLET>
```

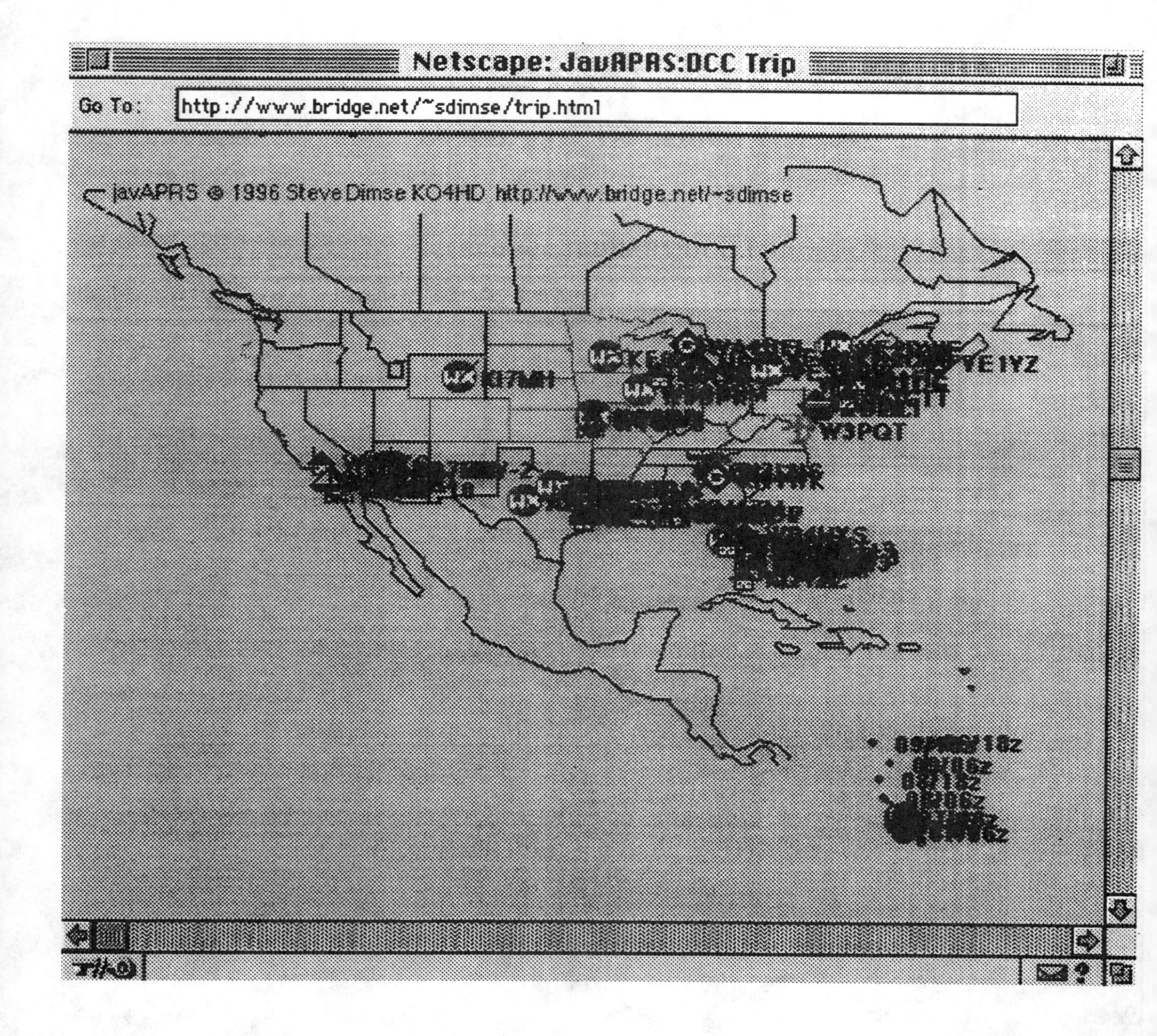

Future directions

The samples above are static data, and although there is some degree of animation while the data is plotted, they could be done just as easily with screen captures and GIF files. Soon I hope to have the system able to access real time data over the net. This is where javAPRS can fill a unique role. The possibility exists to incorporate internet connectivity directly into the standalone APRS programs currently in use, allowing web users to see the data obtained at various APRS stations across the country in real time. Also, it would be possible to write a server program, that could be connected to many different APRS sites and share their data, creating a nationwide APRS network, no longer limited to 300 baud and the vagaries of HF propagation. I plan to pursue these plans, and hope to have some results to share by the 1997 DCC, if not sooner.

Editors Note:

For current information on Java try these sites:

http://www.bridge.net/~sdimse/javAPRS.html

http://www.bridge.net/~sdimse/javAPRSprog.html

On-air Measurements of MIL-STD-188-141A ALE Data Text Message Throughput over Short Links

Ken Wickwire
kwick@mitre.org

For the past six months a colleague W1IMM and I have conducted automated measurements of throughput when ASCII text files are sent over short or "tactical" HF paths using the ALE Data Text Message (DTM) engineering orderwire (EOW). This is a preview of our results.

Tactical HF links generally use either surfacewave or near-vertical-incidence skywave (NVIS). Surfacewave works out to about 50 miles and NVIS to about 300 miles. Multipath, D-layer absorption and interference usually affect NVIS, which usually has lower throughput than communications over "standard" (i.e., long) one-hop skywave paths.

ALE systems' ability to measure channel quality, and use it to make choices of good channels, now offers improved performance over tactical paths. ALE systems employ a slow but robust waveform that uses interleaving and two kinds of forward error correction (FEC) to combat the fading, noise and interference of HF channels.

ALE standards prescribe three engineering orderwire protocols for half-duplex data transfer: the Automatic Message Display (AMD) mode, the Data Text Message (DTM) mode and the Data Block Message (DBM) mode. DTMs can transfer ASCII text using the ALE waveform and an ARQ protocol.

We have carried out more than 200 measurements of throughput (in char/s and char/s/Hz) using DTMs on a 35-mile path. We have used 125-watt Harris RF5022 ALE radios, which implement DTMs of constant size (300 bytes) and "memory ARQ," in which up to six erroneous repeats of a message segment sent as a DTM are stored and compared when necessary in an attempt to construct an error-free segment. The RF5022's ALE firmware segments messages longer than 300 bytes into 300-byte DTMs. These DTM-segments are ACKed one at a time. Experiments suggest that messages 300 to 1000 characters long produce the highest throughput consistent with shortest run time.

Our two stations used broadband sloping longwires that allowed the radios to drive the antennas without tuners. The antennas have both vertical and horizontal components, so that they can launch both surfacewave and NVIS signals.

The ALE modems were programmed to try frequencies between 2 and 16 MHz. (IONCAP runs suggested that any link above 8 or 9 MHz probably used surfacewaves, which were chosen frequently at night, when interference was heavy.) The tests covered seven months from September, 1995, when the average sunspot number was near the bottom of its cycle.

Our measurements were automated by two C-programs. The first runs throughput tests. At the start of a test, and between DTM transfers, the receiving station is scanning the set of programmed frequencies and will stop upon hearing a call. If a link occurs, the calling and receiving stations negotiate a DTM transfer and the sender begins sending the DTM.

The program usually starts by performing a link quality assessment (LQA) exchange, which gives both stations up-to-date info on channel quality. After the exchange, the calling radio links with the receiving one. When the calling radio informs the program that a link has been established, the program commands the local radio to list the channel number and the corresponding LQA scores for later analysis.

When the LQA scores for the linking channel have been logged, the program prepares the local radio to send an orderwire. It then uploads a standard English ASCII text file for transfer. After sending commands that require a particular response, the program invokes a function called `chkresponse()` that scans the serial-port input from the local radio for the appropriate response ("LINKED", "MESSAGE RECEIVED", etc.).

As the program runs, timers set by the computer's clock measure "link time" and "message transfer time." Message transfer time is the time between start of character-by-character uploading of the file to the local radio and receipt of the MESSAGE RECEIVED notification. Transfer time does not include link time. The program calculates throughput by dividing the number of characters sent by the message transfer time. Since the message transfer time includes the few seconds needed for the receiving station to send the MESSAGE RECEIVED frame, the throughput measurements are slightly pessimistic.

The throughput-measuring program writes its results to an archive file. The archive file stores appended, time-stamped data in abbreviated format that can be analyzed off-line by a statistics program. Here is abbreviated statistical output for all the tests run up to 6 May 1996:

```
no_ALE_links = 235
E(link_time_ALE) = 25.96 s, sd(link_time_ALE) = 21.42 s
E(transfer_time_ALE) = 139.9 s, sd(transfer_time_ALE) = 98.7 s
E(no_file_chars_ALE) =  770.4, sd(no_file_chars_ALE) = 615.1
E(tput_ALE) = 5.21 cps, sd(tput_ALE) = 1.16 cps, sd(mean_tput_ALE) = 0.08 cps
max_thruput_ALE = 6.60 cps, E(thruput_ALE/Hz) = 0.003 cps/Hz

Linking histogram:
Channel 1  (2.394 MHz):    1  link
Channel 2  (2.824 MHz):    70 links
Channel 3  (3.166 MHz):    29 links
Channel 4  (4.565 MHz):    8  links
Channel 5  (5.031 MHz):    56 links
Channel 6  (6.870 MHz):    4  links
Channel 8  (7.850 MHz):    1  link
Channel 9  (9.305 MHz):    7  links
Channel 10 (10.330 MHz): 3  links
Channel 11 (10.523 MHz): 5  links
Channel 12 (13.692 MHz): 45 links
Channel 13 (15.487 MHz): 6  links
```

E() and sd() stand for the expectation (average) and standard deviation of a measurement. About two-thirds of a set of measurements will be within one standard deviation of their mean and over 90% will be within two. The `sd(mean_tput_ALE)` here suggests that our sample sizes are big enough to give us high confidence that if we collected more throughput measurements under roughly the same conditions, we would not get average throughputs that differed from the one above by more than a tenth of a character per second. One should keep in mind that our "conditions" correspond to winter and spring operations at low sunspot numbers. Average throughput in other seasons and at much higher sunspot numbers will probably be different.

To calculate the average throughputs per Hertz [`E(thruput_ALE/Hz)`], we divided the average throughput by the ALE signaling bandwidth. For the latter we used the formula for "necessary telegraphy bandwidth" (from the 1992 Dept. of Commerce *RF Management Handbook*):

$$BW = R / \log_2(N_T) + f_{max} - f_{min},$$

where R (= 375 bits/s) is the channel rate, N_T (= 8) is the number of MFSK tones in the ALE waveform, f_{max} (= 2500 Hz) is the highest tone and f_{min} (= 750 Hz) is the lowest tone. For these values, the signaling bandwidth is 1875 Hz. At the end of its output, the statistics program prints histogram values of the frequencies chosen for linking by the sending ALE radio (more on these below).

Because of the relatively low channel rate used by the ALE waveform (375 bits/s or 125 symbols/s), and the high overhead used to provide the waveform's robust forward error correction (about five-sixths of the channel rate), average DTM throughput (about 5 char/s) is modest. What distinguishes ALE DTM transfers from ASCII transfers using several other ARQ protocols, like the AMTOR, PacTOR, GTOR and AX.25 packet protocols, is the fact that on tactical links, DTM transfers are much more frequently successful, especially at night. The main reason for this is ALE's ability to look for another frequency for linking if the previously tried frequency fails. File transfers were successful on roughly 90% of the automated attempts.

Our ALE systems have probably often linked on surfacewave and possibly E-layer frequencies rather than on NVIS frequencies, which lie near the bottom of the HF band. Surfacewave frequencies appear to have been chosen often at night. (Most traditional propagation prediction programs do not suggest surfacewave frequencies for night time or any other operation.) One reason for avoiding NVIS communications at night is that there is almost always more interference on NVIS frequencies at night than during the day. For short-range communications, it is important to use antennas that can launch both NVIS and surfacewave signals; that is, antennas with both vertical and horizontal components.

The relatively small standard deviation of DTM throughput reflects the DTM protocol's restricted ability to adapt to changing conditions. The low variability of throughput over short paths and the reliability of transfer are also reflected in the fairly small difference between average and maximum (6.6 cps) observed throughput.

Average link time was about 26 seconds, with a standard deviation of about 21 s. This standard deviation implies that establishing a link required at least two attempts fairly often during our tests. (A single successful link handshake takes about 20 seconds.)

The histogram data produced by the statistics program are graphed below. The histogram shows that most DTMs were transferred at 2.824, 3.166, 5.031 and 13.692 Mhz. The middle two frequencies were chosen mostly during the day and probably supported only NVIS. The 2.824 MHz frequency appears to support both surfacewave and NVIS. (2.394 MHz may have had antenna-matching problems that caused its poor performance.) The transfers on 13.692 MHz were mostly at night and were probably by surfacewave. It is unlikely that an inexperienced communicator would have tried 13 MHz for night-time operation over this link.

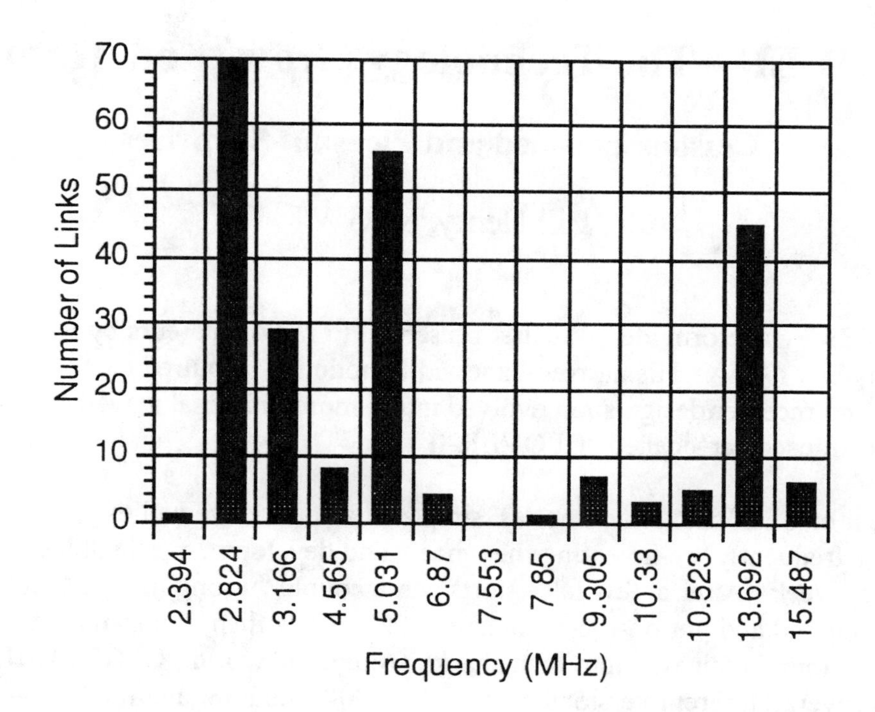

These on-air results suggest that although the ALE DTM engineering orderwire mode is relatively slow, on tactical links it can perform ASCII file transfers more reliably than several other ARQ systems in current use.

Acknowledgments.

I am grateful to John Morgan of the Harris Corporation for providing the Harris RF5022 radio systems with which these measurements were made, and to Bob Levreault (W1IMM) for maintaining one of the ALE stations used for the tests.

CLOVER - The Technology Grows and Matures

Lessons Learned and Pleasant Surprises

Bill Henry, K9GWT

The "CLOVER" waveform idea was first presented to radio amateurs by Ray Petit in the July, 1990 issue of QEX. His narrow-bandwidth mode that required custom radio transmitter and receiver design soon evolved into a more universal waveform for use with any HF SSB transceiver - called "CLOVER-II".

Development of CLOVER-II was not as simple or as fast as we had expected. It seemed like we were frequently "re-inventing the wheel" and development schedules were re-written every week. After a few false starts and a couple "wrong turns", CLOVER-II was finally shipped and put on-the-air in late 1992. Since then, thousands of CLOVER modems have found their way into ham shacks around the world. CLOVER-II is now available in several different versions of the PCI-4000, the P38, and the DSP-4100 modems as well as in customized systems for commercial and government customers. CLOVER has now been used in every conceivable HF radio application. This includes ham radio, ship traffic, aircraft communications, bank data transfer, computer file transfer, image transmission, and even digital voice. Virtually anything that can be digitized and stored in a PC has been sent somewhere via HF radio and CLOVER, often at data throughput rates that match or exceed those we see on 1200 baud VHF packet radio.

Early CLOVER Lessons:

CLOVER started with a great burst of enthusiasm. We had a bunch of new ideas and a "hundred or so" ways to implement each one. The new DSP architecture promised unheard of design freedom. This was our first product where virtually *every detail* was set by software. We soon became quite familiar with the phrases: "It's only software." - usually followed by - "It will take *how long* to make that change?" We demonstrated the "gas laws" often, especially the one that states "gas expands to fill the available volume". Our version is "Software expands to fill available memory space and consume all processor time".

CLOVER itself experienced many evolutional changes. The first version had a bandwidth of 100 Hz, one tone and used only one phase shift modulation mode. During development, the CLOVER waveform soon expanded to four tones, a bandwidth of 500 Hz, and a total of 160 different modulation modes - a real "knob-twister's paradise"! Coming from the HF packet world, we also intended to fix some of the problems we'd seen with AX.25 on 20 Meters. The CLOVER ARQ protocol therefore includes selective block repeat, in-block error correction coding, bi-directional data flow (no "OVER" command), and adaptive modulation control. While some or all of these ideas had been used previously, CLOVER was the first to combine the multi-level, multi-mode modulation waveform with an ARQ protocol specifically designed for use over HF radio links. No doubt about it, CLOVER is a *complicated* mode.

Predicting software completion turned out to be very risky business. The first target release date was "April, 1991". This quickly slipped to "fall, 1991" to "April, 1992" and finally to "November, 1992 (actually the *last* day of November, 1992). AND - we weren't done! During the first 6 months of CLOVER's commercial life, we issued 8 no-cost software up-grades. We've since provided almost 50 software changes to CLOVER and DSP software, the most recent occurring this summer - July, 1996.

While the hardware schematic and block diagram have been very stable, parts cost and availability have been continuing problems. Price and delivery promises made in 1990 by suppliers have proven to be "optimistic dreams". This is particularly true of the Motorola DSP components. The cost of these parts remains high and periodically they become "non-deliverable". In early 1995, frustrated with this high cost and unpredictable delivery, we redesigned the modem to use the much less expensive and more available TMS320C25 processor and related parts. The P38 modem is the result, the first and so far only DSP modem with a list price under $400 (it usually sells in the "mid-$300 range).

Finally, CLOVER has experienced the typical problems of a pioneering mode. Correctly tuning the radio receiver to exactly match a CLOVER signal takes practice. Virtually all radios made since 1990 can meet the ±10 Hz tuning increment requirement, but you need a "soft touch" on the knob and patience to make small corrections and then wait. Once learned, tuning a CLOVER signal is easy - BUT - it's not like any other mode you've ever used! Likewise, tuning the transmitter for no ALC and less than "Max-Smoke" output takes personal discipline, particularly hard for us old-time RTTY types. These are not new problems in HF radio. The early SSB pioneers faced the very same problems in the 1950's. Tuning-in an SSB signal was a lot harder than tuning an AM signal. 40 years later, we think nothing of tuning an SSB signal - "No big deal"! Likewise, we all soon learned that adjusting a linear amplifier was a lot different than tuning the AM finals for "cherry-red plates". Inaccurate tuning and incorrect transmitter adjustment severely limit CLOVER performance and have frequently been the underlying problem behind "it doesn't work" complaints. After 5 years, CLOVER is finally tuning the corner where we are getting comfortable with using it and can also say "no big deal" about these problems.

Marine CLOVER:

Ships at sea have used the "SITOR" mode for HF data communications for 30 years. Special frequency channels are allocated for ship-to-shore *Narrow Bandwidth Direct Printing Telegraph* (NBDP) use. "Paired frequencies" are used with ships transmitting on one set of channels and shore stations on another set. Within the "ship" or "shore" sub-bands, the channels are spaced exactly 500 Hz apart. Particularly in the case of shore station allocations, every 500 Hz wide channel is in active use, usually by very powerful transmitters (1 kW to 10 kW). It is therefore very important that adjacent channel interference be prevented. HAL has teamed with Globe Wireless of Half Moon Bay, California to develop a special version of CLOVER that is tailored to meet ship-to-shore requirements. While the original CLOVER-II waveform has a frequency spectra that is exactly 500 Hz wide (@ -50 dB), this leaves no "guard band" between adjacent channels for tuning error. A marine version was therefore developed that has a -50 dB bandwidth of 400 Hz, well within the FCC Part 80 limits for use on NBDP HF channels. We call this version "CLOVER-400". The spectra of CLOVER-400 and the FCC limits are shown in Figure 1.

Globe Wireless has created a "Global Radio Network" of public coast stations distributed around the world. All public coast stations are tied into the global network via whatever common carrier service is most cost effective for that location (wire line, satellite, Internet, etc.). Regardless of its location at sea, a ship can establish HF radio contact with one or more network coast stations. When not actively sending data, each ship receiver constantly scans the frequency list of Globe stations, listening to traffic or SITOR "free signals", and logging signal quality data for each station heard. At any given time, the shipboard computer therefore "knows" which frequency and station is optimum for communications. Each ship acts as a "passive sounder", obtaining "LQA" data (Link Quality Assessment) for each usable coast station *without the need to transmit.*

All coast stations are equipped to use either SITOR or CLOVER-400 on NBDP HF channels. To maintain compatibility with older vessels, communications start in SITOR mode and then switch to CLOVER-400 when the shore station recognizes a CLOVER-equipped ship. Use of CLOVER not only increases the speed at which data is delivered, it also provides error-corrected 8-bit data transfer of any computer file, be it text, data, image, or executable software. In fact, new software for the CLOVER modem on each ship is passed via CLOVER on HF radio - the system upgrades itself. All this is accomplished at a fraction of the cost the user would otherwise have to pay for satellite communications.

Voice Bandwidth CLOVER:

While amateur data modes emphasize narrow bandwidth (500 Hz or less) to conserve limited spectrum, commercial and military HF allocations are virtually all for "voice bandwidth" channels (ship-to-shore excepted). The U.S. Civil Air Patrol (CAP) has made full use of the four "tone channels" of CLOVER-II to expand the capacity of their limited number of HF channel allocations. The CAP technique is to share an HF voice channel between four non-interfering and independent CLOVER ARQ links. This application takes advantage of CLOVER's exceptionally high stop-band suppression. Selection of the CLOVER-II "tone channel" is a user-set feature included in all PCI-4000 and DSP-4100 modems (not available in the P38 modem). The CAP CLOVER "multiplex" of a voice channel is shown in Figure 2.

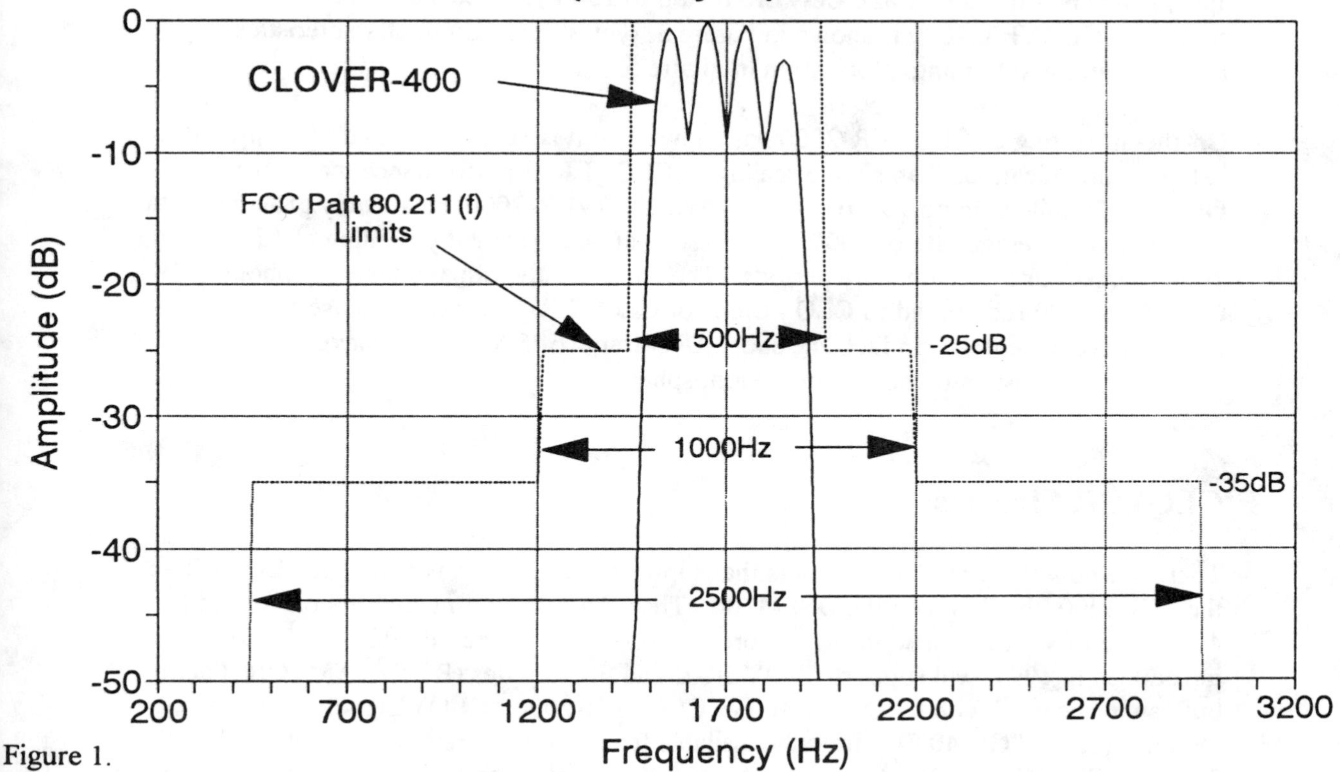

Figure 1.

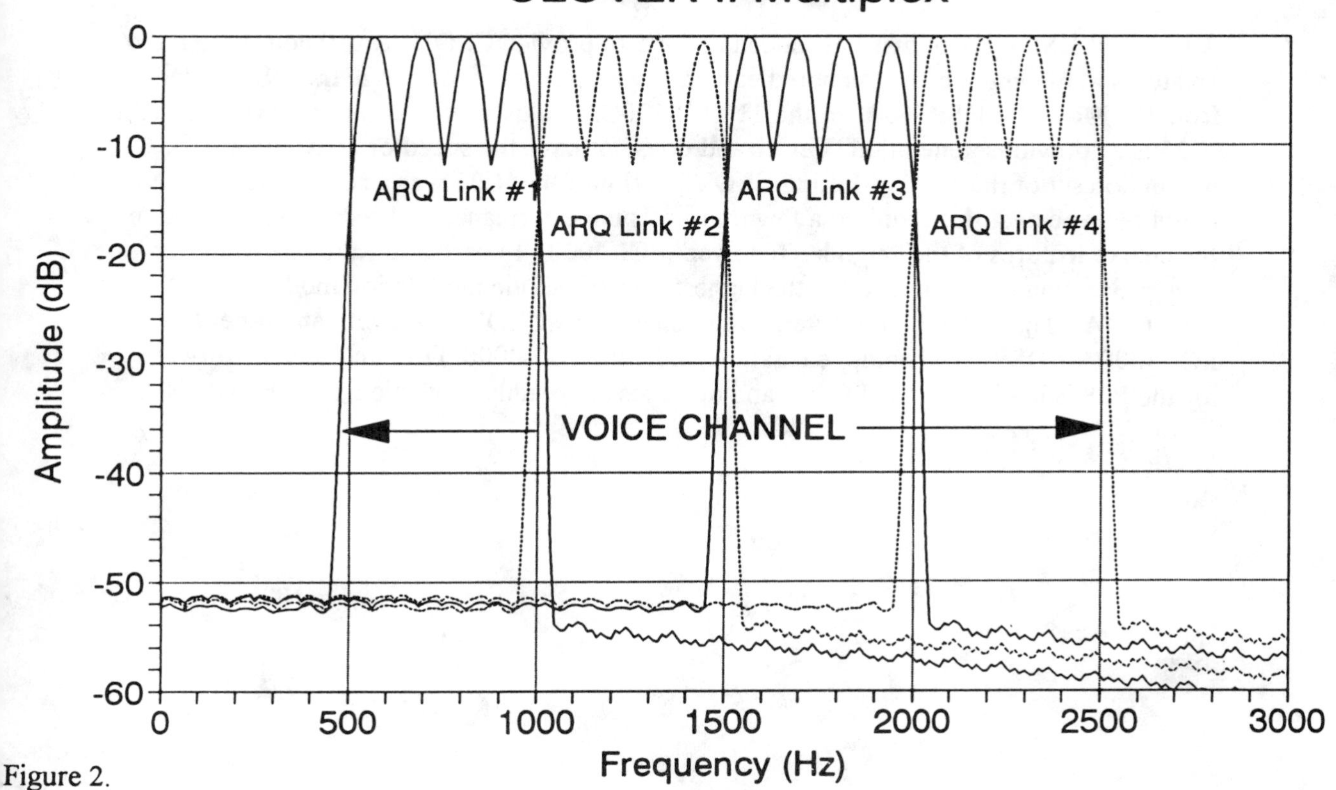

Figure 2.

CLOVER technology has also been expanded to gain higher data throughput by using more tones and a higher symbol rate. "CLOVER-2000" uses 8 tones, a symbol rate of 62.5 Baud, occupies a bandwidth of 2000 Hz, and has a data throughput rate that is approximately 4 times that of CLOVER-II - up to 250 bytes/sec. (2000 bits/sec). The spectra of CLOVER-2000 is shown in Figure 3, typical throughput characteristics in Figure 4, and ARQ timing information in Figure 5.

On-the-air testing of CLOVER-2000 has shown that it is very robust and often provides better communications than simple scaling of CLOVER-II performance might predict. Over an 800 mile path on 10 MHz at mid-day, CLOVER-2000 has reliably provided data transfer at an average rate of 1000 bits-per-second, and frequently running at 1500 to 2000 bits/sec for 5 to 10 minute periods. The wider bandwidth and faster symbol rate of CLOVER-2000 (compared to CLOVER-II or CLOVER-400) do not cause excessive block failure or repeats. In fact, the short ARQ frame of 5.5 seconds makes CLOVER-2000 very responsive to changes in the ionosphere.

CLOVER Modems:

There are now three modem products that support one or more versions of CLOVER - the PCI-4000, the P38, and the DSP-4100. There have been 4 major versions of the PCI-4000, two of which are still in current production. The "standard" PCI-4000, first introduced in 1993, will support CLOVER-II and FSK modes (RTTY, AMTOR, Pactor) but cannot be used with CLOVER-400 or CLOVER-2000. CLOVER-400 capable modems (called "GL-4000") are only available from Globe Wireless. Use of CLOVER-2000 (or CLOVER-400) requires a faster 68000 processor and additional memory. This hardware version is called the "PCI-4000 Plus". All versions of the PCI-4000 use on-board DSP and 68000 software that is upload via the PC data bus. Software up-dates can be quickly and easily obtained from the TECHLINE BBS.

A low cost DSP modem was introduced in 1995 to provide CLOVER technology to radio amateurs at minimum cost. As noted earlier, it was necessary to change the DSP system from the Motorola DSP56001 to the TI TMS320C25 (plus related components). The TI 'C25 is a "previous generation" device and does not have the speed or convenient instruction set of the 56001. Fitting CLOVER-II into the 'C25 was a real "squeeze". As might be expected, this required a few trade-offs in performance and reduction of some of the deluxe features of the "high horsepower" PCI-4000. In particular, the P38 does not have processing power to receive the highest two amplitude modulation modes - 8P2A and 16P4A. The P38 will, however, communicate via CLOVER-II with any other P38, PCI-4000, or DSP-4100 equipped station. Like the PCI-4000, DSP and 68000 software for the P38 is loaded via the PC bus and up-dates are readily available from TECHLINE.

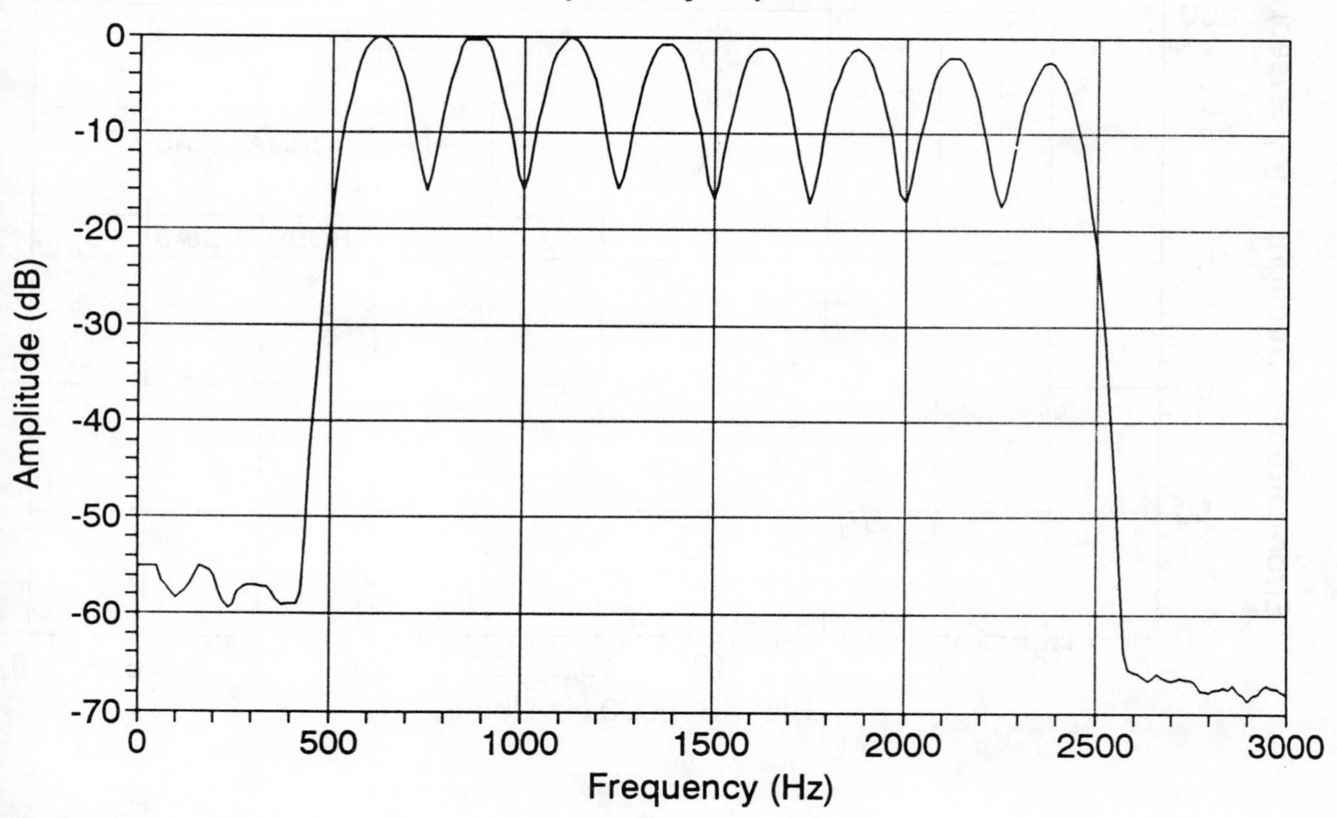

CLOVER-2000
Frequency Spectra
Amplitude (dB)
0
-10
-20
-30
-40
-50
-60
-70
0
500
1000
1500
2000
2500
3000
Frequency (Hz)

Figure 3.

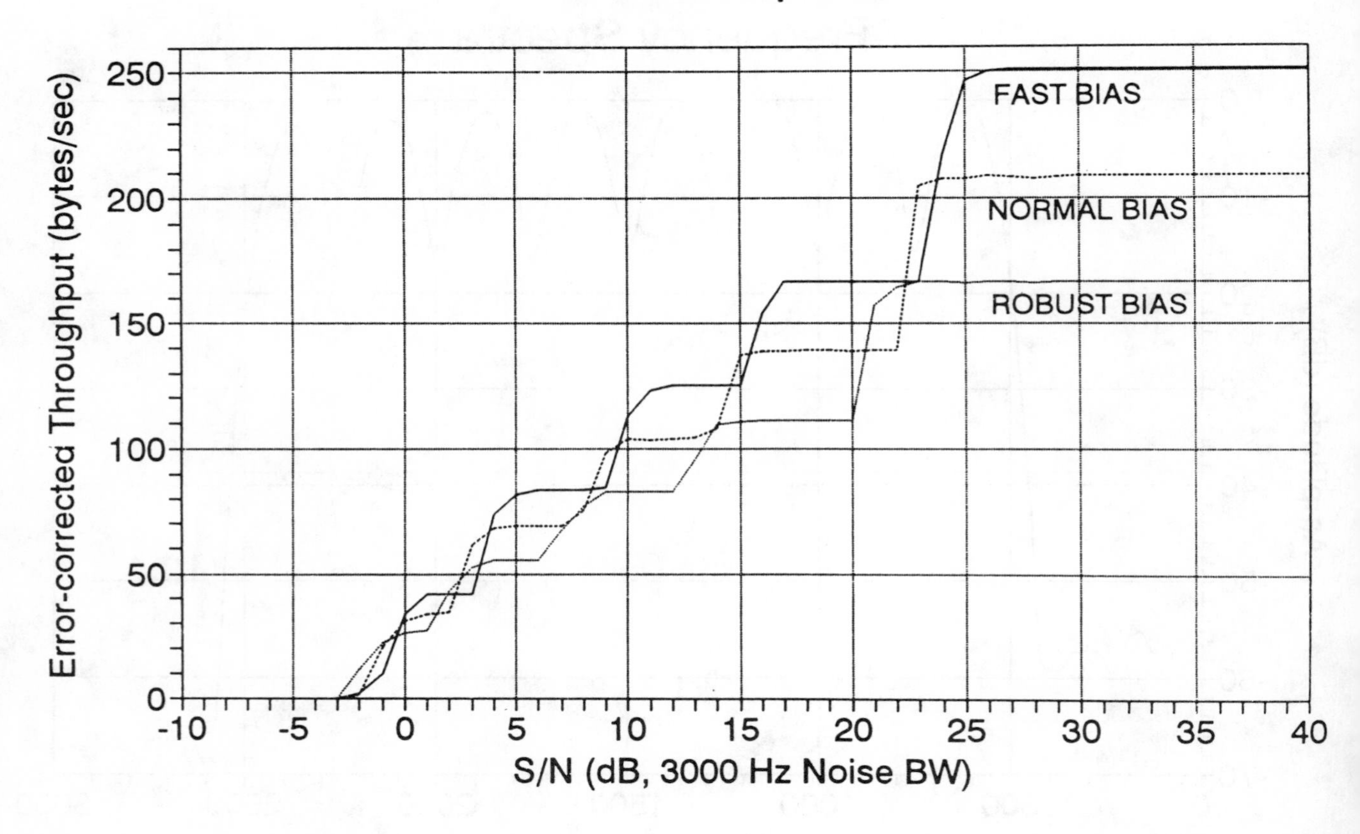

CLOVER-2000 THROUGHPUT
Bias Comparison
Error-corrected Throughput (bytes/sec)
250
200
150
100
50
0
-10
-5
0
5
10
15
20
25
30
35
40
S/N (dB, 3000 Hz Noise BW)
FAST BIAS
NORMAL BIAS
ROBUST BIAS

Figure 4.

CLOVER-2000 MULTI-BLOCK ARQ DATA FRAME

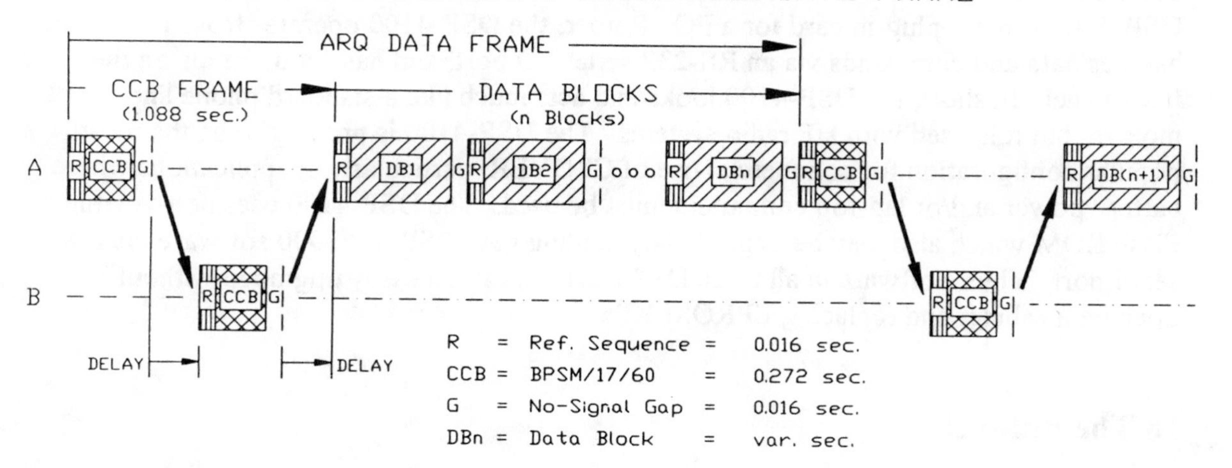

ROBUST

ROBUST BIAS (60%) RATE	MOD	BLOCK	BYTES/FRAME	MAX ERRORS	BLOCK TIME	BLKS/FRAME	ARQ FRAME TIME	THRU-PUT BYTES/SEC
165	16P4A	255	900	300	0.688 sec	6	5.440 sec	165.4
110	8P2A	255	600	200	1.024 sec	4	5.440 sec	110.3
83	8PSM	255	450	150	1.360 sec	3	5.440 sec	82.7
55	QPSM	255	300	100	2.048 sec	2	5.440 sec	55.1
28	BPSM	255	150	50	4.080 sec	1	5.440 sec	27.6

NORMAL

NORMAL BIAS (75%) RATE	MOD	BLOCK	BYTES/FRAME	MAX ERRORS	BLOCK TIME	BLKS/FRAME	ARQ FRAME TIME	THRU-PUT BYTES/SEC
207	16P4A	255	1128	186	0.688 sec	6	5.440 sec	207.4
138	8P2A	255	752	124	1.024 sec	4	5.440 sec	138.2
104	8PSM	255	564	93	1.360 sec	3	5.440 sec	103.7
69	QPSM	255	376	62	2.048 sec	2	5.440 sec	69.1
35	BPSM	255	188	31	4.080 sec	1	5.440 sec	34.6

FAST

FAST BIAS (90%) RATE	MOD	BLOCK	BYTES/FRAME	MAX ERRORS	BLOCK TIME	BLKS/FRAME	ARQ FRAME TIME	THRU-PUT BYTES/SEC
249	16P4A	255	1356	72	0.688 sec	6	5.440 sec	249.3
166	8P2A	255	904	48	1.024 sec	4	5.440 sec	166.2
125	8PSM	255	678	36	1.360 sec	3	5.440 sec	124.6
83	QPSM	255	452	24	2.048 sec	2	5.440 sec	83.1
42	BPSM	255	226	12	4.080 sec	1	5.440 sec	41.5

Figure 5.

The DSP-4100 modem was introduced in late 1995. Unlike the PCI-4000 and P38, the DSP-4100 is not a plug-in card for a PC. Rather, the DSP-4100 operates from 12 VDC, handles data and commands via an RS-232 serial I/O port, and has status lamps on the front panel. In short, the DSP-4100 looks and acts much like a standard phone line modem, but it is used with HF radio systems. The DSP-4100 is proving to be the most popular configuration for commercial use of CLOVER, particularly in applications where battery power and/or lap-top computers must be used. The DSP-4100 uses non-volatile Flash ROM which also may be upgraded by loading new DSP or 68000 software via the serial port. Thus, software in all three DSP modems can be easily upgraded without opening a cabinet and replacing EPROM IC's.

In The Future:

CLOVER continues to grow and improve. Present software gives performance and features that were not available when CLOVER started. The "it's only software" development problem is finally working in our favor. Within limits, virtually every CLOVER modem sold can be upgraded in the field at little cost to the user. Custom versions of the hardware, terminal software, and CLOVER itself have been and continue to be developed to meet the needs of its users. CLOVER-2000 and the DSP-4100 have greatly extended the horizon for use of CLOVER technology. CLOVER modem hardware and software is already being installed inside some models of HF transceivers. In these systems, there are no outward signs that CLOVER is being used unless you turn up the receiver volume control and hear the distinctive 4- or 8-tone "CLOVER twitter".

The future direction of CLOVER technology - waveform, protocol, software, and hardware - will be determined by the user. CLOVER provides the "missing link" that makes HF radio an attractive cost-effective alternative to satellite or wire-line data communications.

CLOVER BIBLIOGRAPHY

Townsend, Jay: "The HAL Communications P38 HF Radio DSP Modem" (New Product Review), CQ, October, 1995, p. 30, CQ Publishing, Hicksville, NY.

Vinson, Glen: "HAL P38 vs AEA PK-232", Digital Journal, September, 1995, pp 12 & 13, International Digital Radio Association ("IRDA"; formerly American Digital Radio Society - "ADRS"), Box 2550, Goldenrod, FL. 32733-2550.

Ford, Steve: "Product Review - HAL Communications P38 HF Modem", QST, August, 1995, pp 71-73, American Radio Relay League (ARRL), Newington, CT.

Blegen, Hal: "The HAL P38 DSP", Digital Journal (RTTY Journal), August, 1995, pp 8 & 22, International Digital Radio Association ("IRDA"; formerly American Digital Radio Society - "ADRS"), Box 2550, Goldenrod, FL. 32733-2550.

Schulz, Fred: "P38, die neue CLOVER-Platine" SWISS ARTG Bulletin (6/95), pp 14-17, SWISS Amateur Radio Teleprinter Group, c/o Arturo Dielet, HB9MIR (Secretary), Blauenwig 8, 4335 Laufenburg, Switzerland (in German).

Walder, Stephan: "Neu bei der SWISS-ARTG: P38 von HAL", SWISS ARTG Bulletin (4/95), pp 8 & 13, SWISS Amateur Radio Teleprinter Group, c/o Arturo Dielet, HB9MIR (Secretary), Blauenwig 8, 4335 Laufenburg, Switzerland (in German).

Hollingworth, Jack; "The New HF Data Mode: CLOVER II", RADIO COMMUNICATIONS, November, 1993 (pp 52-54), December, 1993 (pp 62-64), January, 1994 (pp 68-70), RSGB, Lambda House, Cranborne Road, Potters Bar, Herts. EN63JE England

Schilling, H.J., and Paul Williams: "Clover ein neus digitales Funk-Verfahren fur Kurwelle", CQ DL, June, 1993, pp 385-387, Deutschen Amateur-Radio-Clubs (DARC), Postfach 11 55, 3507 Baunatal, Germany (in German).

Healy, James W.: "HAL Communications PCI-4000 CLOVER-II Data Controller" (Product Review), QST, May, 1993, pp 71-73, American Radio Relay League (ARRL), Newington, CT.

Townsend, Jay: "CLOVER - PCI-4000", RTTY Journal, April, 1993 (pp 3-4) May/June, 1993 (p 20), 1904 Carolton Lane, Fallbrook, CA. 92028-4614.

Walder, Stephan: "CLOVER - Eine neue Betriebsart", SWISS ARTG Bulletin (1/93), pp 5-12, SWISS Amateur Radio Teleprinter Group, c/o Arturo Dietlet, HB9MIR (Secretary), Blauenwig 8, 4335 Laufenburg, Switzerland (in German).

Schulz, Fred: "CLOVER ist eingetroffen", SWISS ARTG Bulletin (1/93), pp 13-15, SWISS Amateur Radio Teleprinter Group, c/o Arturo Dietlet, HB9MIR (Secretary), Blauenwig 8, 4335 Laufenburg, Switzerland (in German).

HAL Communications, PCI-4000 CLOVER-II Data Modem Reference Manual and PC-CLOVER Operator's Manual, November, 1992, HAL Communications Corp., Urbana, IL ($25.00).

Henry, George W., Ray C. Petit: "CLOVER - Transmision de datos rapida en HF", CQ (Spanish Language edition), November., 1992, pp 45-48, CQ Publishing, Hicksville, NY (in Spanish).

Petit, Ray C.: "The CLOVER-II Communication Protocol - Technical Overview", ARRL 11th Computer Networking Conference Proceedings (October, 1992), American Radio Relay League (ARRL), Newington, CT.

Henry, George W., Ray C. Petit: "HF Radio Data Communications: CW to CLOVER", Communications Quarterly, Spring, 1992, pp 11-24; CQ Publishing, Hicksville, NY.

Henry, George W., Ray C. Petit: "CLOVER: Fast Data on HF Radio", CQ, May, 1992, pp. 40-44; CQ Publishing, Hicksville, NY.

Horzepa, Stan (ed); George W. Henry & Ray C. Petit: "CLOVER Development Continues", "Gateway", QEX, March, 1992, pp. 12-14, American Radio Relay League (ARRL), Newington, CT.

Henry, George W., Ray C. Petit: "CLOVER Status Report", RTTY Journal, January, 1992, pp. 8-9, Fountain Valley, CA.

Henry, George W., Ray C. Petit: "Digital Communications for HF Radio - AMTOR & CLOVER"; paper presented at Amateur Radio Digital Communication Seminar, St. Louis, Mo., October 26, 1991.

Petit, Ray C.: "CLOVER-II: A Technical Overview", ARRL 10th Computer Networking Conference Proceedings (1991), pp. 125-129; American Radio Relay League (ARRL), Newington, CT.

Petit, Ray C.: "CLOVER is Here", RTTY Journal, Fountain Valley, CA; January, 1991, pp. 16-18; February, 1991, pp. 12-13; March, 1991, pp. 16-17; April, 1991, p 10.

Petit, Ray C.: "The CLOVERLEAF Performance-Oriented HF Data Communication System", QEX, July, 1990, pp. 9-12; reprinted in ARRL/CRRL 9th Computer Networking Conference Proceedings (1990), pp. 191-194; American Radio Relay League (ARRL), Newington, CT.

Chapter 3
Networking

Conducted By Stan Horzepa, WA1LOU
One Glen Ave, Wolcott, CT 06716-1442
Internet: stan.horzepa@corp.gdc.com CompuServe: 70645,247

RMNC/FlexNet: The Network of Choice in Western Europe

Eric Bertrem, F5PJE, has read articles in Packet Perspective dealing with NET/ROM, TCP/IP, TexNet, ROSE and other networking systems used in the US, but has never read anything here about RMNC/FlexNet, *which is used throughout Europe. This month Eric describes how RMNC/FlexNet works and how to use it.*

The Hardware

The Rhein Main Network Controller (RMNC) is the hardware portion of the system and FlexNet is the software portion; 95% of the software is written in C and the remainder, the low-level I/O, is written in 6809 assembly language. Development of the system began in Germany in 1987 and came to fruition when the first RMNC/FlexNet digipeater came on the air in 1988.

Typically, the controller uses the following configuration of equipment:

Reset II card that provides I/O for remote controls, clock signals for the other cards and resets for card failures.

Solomaster card for control of and uploading software to the *Slave* cards.

Slave cards, one for each radio link, that run up to 115,200 bauds using 4, 8, 12 or 16-MHz CPU clocks. The latest version, RMNC 3, includes the following options: a KISS/RS-232 interface to link the digipeater to any PBBS capable of using KISS, a 1200-baud modem with digital echo of the received frames (similar to the bit regenerator option of TAPR 9600 modems) and 9600-baud FSK DF9IC (G3RUH-compatible) modem with digital echo, if needed.

*ANET*s are fully compatible French versions of the RMNC cards that were developed by F5CAU and F6BNY. Each *ANET*s card generates its own clock signals, therefore, the *Reset II* is unnecessary unless you need its I/O port.

All the cards plug into the bus card (16 cards maximum at each digipeater) and can be connected easily to external modems operating from 4800 to 38,400 bauds (many modems are available in Germany for about $90 each). Using such high-speed modems means using dedicated transceivers. The first dedicated links in Germany began on 70 cm, but the band quickly became overcrowded, so DF9IC designed a 23-cm half- and full-duplex transceiver, *Interlink I* and *Interlink II*, respectively. They are FSK-ready with 1 to 20 watts output.

Today, most German links are on 1.2 GHz at 9600 or 19,200 bauds. The network is efficient and provides fast response to downloads, uploads and even keyboard-to-keyboard real-time chats. I have a QSO every day with a friend 700 km from home through seven digipeaters and often download new versions of German software from a PBBS 1000 km from here with no link failures.

The Software

First used in Germany, FlexNet quickly spread into Switzerland and then to France, where we previously had unreliable networks with slow (1200-baud) links. FlexNet is also in operation in Austria, Belgium, Hungary and Holland. Manuals are available in French, German and, soon, English. Here's how it works (the following examples have been translated into English).

After I connect to my local RMNC/FlexNet digipeater, I receive the following message:

(1) CONNECTED to F6BIG-7 RMNC/FlexNet V3.3a
***Mt Semnoz (1700m) JN35BT 17 km SW Annecy (74)**
(H)elp = commands - (L)inks - (A) = news - (Q)uit
*** 'c f6big-8' => bbs Annecy ***
*** 'c hb9iac-8' => dx-cluster Geneva ***
Please, only ONE connect at a time, thanks to all the users!
Users access on 70 cm = 433.775 MHz
This channel will change to 9600-baud G3RUH in the near future. Get the equipment!

Next, I invoke the P (for Parameters) command to get the pertinent parameters of the digipeater. In response, the digipeater sends me the following information for each of its ports:

po id td qso usr tifr rifr txkby rxkby qty mode links

"po" is the port number (one *Slave* card per port)

"id" the SSID used (no SSID means the port is linked to another digipeater)

"qso" and "usr" represent the number of QSOs and users on the port.

"tifr" and "rifr" are the number of I frames sent and received within the last 10 minutes.

"txkby" and "rxkby" are the number of kilobytes sent and received during the last 10 minutes.

"qty" refers to the quality of the link.

"mode" indicates the speed of each port and the CPU clock used by the card on that port (+ for 8 MHz, ! for 12 MHz, # for 16 MHz).

Other parameters include "d" for full-duplex, "t" for external transmit clock, "r" for external receive clock and "z" for the NRZ mode. The time to reach a neighboring digipeater (in 100 ms steps) is provided.

On FlexNet, user ports have SSIDs (for example, SSIDs 2, 7 and 6 on F6BIG-2). These help remote stations to choose the correct "output" on each node, in conjunction with the MH function which records the last 200 stations heard on each port. If my intended connection is recorded in the destinations list, that record is used to make the connection.

Next, I invoke the D (for Destinations) command:

D DBØZDF

This asks the system if DBØZDF is a known destination, that is, is DBØZDF recorded in the destinations list? If it is, I'll receive the available user SSIDs and time needed to reach it (the T parameter). The system responds with

***** DBØZDF (0-12) T = 319**

which indicates that DBØZDF is a known destination, so I don't have to specify the path to connect with DBØZDF; all it takes is C DBØZDF and the system responds with

link setup...
***** route: HB9X HB9EAS F6KDL DBØORT DBØCPU DBØAAI DBØZDF**
***** connected to DBØZDF**
RMNC/FlexNet V3.3c Duplex-Digipeater DBØZDF Mainz JN49cx.
QRG R 64 Siehe
A = News C = Conference mode
I = Information H = Help

The system's response to my P (Parameters) command indicates that this digipeater is one of the best in Germany. Run by many hams on the RMNC/FlexNet team, speeds here are high (up to 19,200 bauds), using *Interlink* 23-cm transceivers and—most of the time—antenna dishes to improve link quality without needing much power.

Time to Disconnect

This was only a quick tour of FlexNet. There are many other user and SysOp commands available, and the system is simple and its links are quite reliable.

Recently, the RMNC/FlexNet team developed a new concept called "PC/FlexNet." The software is the same, but the hardware is simply any 386-based computer with a minimum of 540 kilobytes of RAM and a 10-megabyte hard disk. A BayCom USCC plug-in card provides two, four or eight radio ports. PC/FlexNet also supports Ethernet cards, BayCom modems, KISS links and soon all the other SCC cards. In the near future, it will have an IP router and a "ping-pong" conference mode (a network that allows you to chat with friends far away from your location by simply connecting to your local digipeater).

If you need more information or you have any questions, please contact me via packet at F5PJE @ F6BIG.FRHA.FRA.EU or via the Internet at f5pje@stdin.gatelink.fr.net

AVAILABILITY OF SEVENTY 9600 BAUD PACKET CHANNELS ON TWO METERS

By Bob Bruninga WB4APR @ WB3V.MD

Unbelievable? Not really! With the advent of the latest 9600 baud packet radio modems, there is an un-exploited mechanism for opening up dozens of half duplex data channels without ANY impact on existing voice and data bandplans. Read this proposal thoroughly before jumping to any conclusions. First you must consider two apparently un-related facts:

TINY PACKETS: With the new 9600 baud packet modems, not only is there potential for higher speeds, but the packets are also 8 times shorter than conventional packets. This means that each packet occupies the channel for less than 1/10 of a second at a time. If another signal appears on the data channel, then the presence of the new signal can be detected in 0.1 second.

AVAILABLE CHANNELS: Throughout the two meter band, in every corner of the country, there are almost 70 FM channels assigned to single VOICE receivers that, in general, only use their single frequency about 1% to 40% of the time. If a mechanism could be designed to permit these single receivers to continue to use their frequencies on a primary basis, at any time, with priority access and control, then the rest of the time (60% to 99%) they could use their channel for moving digitial data. This mechanism could more than double the bandwidth available in the two meter band! (I used 2 meters only to get your attention. One hundred twenty channels on the 440 band is the actual target because of the availability of 2 Watt 9600 baud UHF data radios for under $150 each.)

VOICE/DATA CHANNEL SHARING: In the past, attempts to share voice and packet have all failed, not because it is a bad idea, but because it has not been done properly. Under a very unique set of conditions, however, voice and data can easily share a narrowband FM channel, IF:

1) VOICE has priority at ALL TIMES and all voice operations are completely transparent to the voice user
2) DATA can never interrupt or attempt to use a busy channel
3) VOICE users can pre-empt/interrupt data instantly at ANY time
4) VOICE users do NOT hear packets or in any way are encumbered by shared use.
5) NO MODIFICATIONS TO ANY voice radio is required

In other words, voice users do not even know that the channel is shared with data. To make data use of a channel work on a secondary basis (behind the scenes) with NO impact on voice usage, there are additional requirements:

1) There must be ONLY one voice RECEIVER on the channel, and it MUST be able to hear EVERY voice user on the channel.
2) This single voice receiver and its conventional COR or un-modified squelch circuit must have total control over channel use.

THE BEST KEPT SECRET: Think for a moment about the input frequency of a typical voice repeater. There is only one receiver listening, and it can hear EVERYONE that desires to use the repeater. If the repeater RECEIVER does not hear anyone using the channel, then IT alone can decide to use the channel itself for data! Now if the data is transmitted in 0.1 second bursts, FROM THIS SITE (on the input channel), with a pause in-between to listen for voice users, then no one will be denied access to the voice repeater for any longer than 0.1 second! Also, while the repeater is transmitting data on its input frequency, nothing is being transmitted to anoy users on the output! In this manner, we can TRANSMIT data FROM this repeater site at 9600 baud on the input frequency WHENEVER THE REPEATER IS NOT OTHERWISE BEING USED FOR VOICE.

MAKING A DATA LINK: Now combine one such voice repeater with another operating in the same manner, and you have a two-way data link between these two sites. Not only is this a FULL DUPLEX channel, but it also operates with NO HIDDEN transmitters, and NO CONTENTION, because there is only ONE transmitter on each such channel. From a data perspective, each repeater site is a data node that transmits on ONE assigned frequency and can LISTEN on as many additional channels as desired. This is an ideal multi-node backbone network for passing traffic point-to-point over long distances!

HARDWARE: To avoid even momentary delays during normal repeater conversations, the external carrier detect of the modem is not only driven by the COR of the repeater receiver, but also by the HANG-TIME of the repeater transmitter as an indication that the repeater is engaged in a conversation. This way, data will only be transferred after the repeater has been unused by voice users for a while.

FORWARDING: Obviously, this type of channel will NOT support any traffic where an impatient human is on one end. This channel, however, is ideal for off-hour BULK forwarding between multi-channel level-4 nodes under computer control.

BAUDRATE: Although I suggested off the shelf 9600 baud radios, the use of 15 KHz repeater channels in most areas of the country might require slowing to 4800 baud to be certain that all energy is contained within 12 KHz. Also, the $139 TEKK 2 watt UHF data radio was actually designed commercially for 4800 baud, and performs excsllently at that rate. Also, to avoid co- channel interference, operating at power levels above 2 watts is probably not advisable. Fortunately, 9600 baud data sounds just like white noise, and usually will NOT open the squelch of another repeater anyway. Finally, to experiment with this concept, you can actually try conventional 1200 baud during off hours. You can limit packets to a single frame (about .8 second instead of the usual 3 to 5 seconds), by setting MAXFRAME to 1 instead of the default 7. By adding an additional hang-time timer, so that the DATA mode is only activated when the repeater is unused for over 3 minutes, a new voice in-the-night will probably not even notice a maximum of 0.8 second delay.

SYSTEM LINKS: Remember, ONLY THE REPEATER ITSELF can transmit on its own INPUT frequency. The way to build a network is then to have each such repeater node TRANSMIT on its own input frequency and similarly LISTEN to the other nodes transmitting on THEIR input frequencies. To get data from your basement BBS, simply have it LISTEN to the repeater node on the input frequency and the BBS then TRANSMITS on its OWN unique frequency. The repeater node then listens on this unique frequency. If you are willing to cut throughput in half in order to save on hardware costs, you can let your BBS transmit (at low power) on the same frequency as some distant repeater node (its voice input freq). This causes a hidden transmitter problem at your repeater which will hear both your BBS and the remote repeater, but with the tremendous bandwidth available, this is probably not a problem. The following diagram shows a typical arrangement for the nodes located at a pair of 146.34/94 and 147.81/21 voice repeaters.

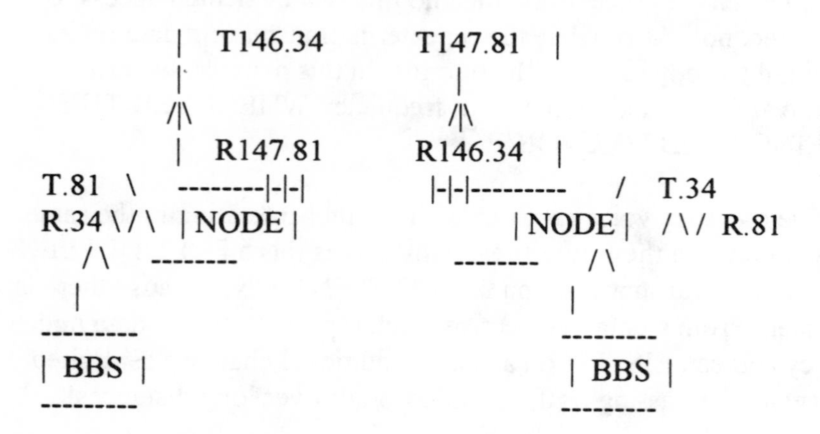

Notice that by pairing up a 146 MHz repeater with a 147 MHz repeater, you get at least 1.6 MHz spacing between the two digital frequencies and about 1 MHz between the digitial and the Voice repeater output (which of course is NEVER transmitting when the digital is in use.

Also notice, that to save dual receivers at the NODE sites, we are cheating a PURE network design by allowing the BBS's to also transmit on the same frequecy as the DISTANT repeater node. This must be done with a beam or low power so that the distant repeater CANNOT hear nor even detect the presence of the BBS. Otherwise, the BBS would KEY UP the distant VOICE repeater!

CONCLUSION: Even if only 30% of the voice repeaters begin to share their channels, this could open up over 600 Kbytes PER SECOND of additional digital forwarding capacity on 2m and 70 cm! Why not?

Building a Packet Network

by Karl Medcalf WK5M

Background

Since the beginning of amateur packet radio, users have tried to push the limits. This has taken many forms: how far can I get, how much data can I pass, how fast can I go.

In 1987, Software 2000, Inc. developed the NET/ROM™ code, which replaced the EPROM in TAPR clone TNCs, in an attempt to improve the packet situation. This code provided the first attempt to build a network using amateur packet radio. Much of the current network system throughout the world is based on this code, and new implementations continue to arrive on the scene.

Purpose

This paper will present various viewpoints on network construction, and does not intend to imply that any one concept is superior to any other. It is intended to provide node operators (current and future) with ideas for consideration to help improve the existing system.

We'll first present the network as it exists in many areas, mostly at 1200 baud on both user ports and backbone ports, and then present ideas for expanding and modifying the existing systems to provide higher throughput and reliability.

Network Concepts

Packet radio networks are intended to provide users with the necessary means to pass data from point A to point B with little regard for the details involved. As such the network node operators must be aware of the impact that any changes and additions to the network may cause.

When amateur packet radio began, each packet station was virtually an independent entity. Users could connect to nearby stations and pass data unencumbered, but long distances could not be spanned easily. Every packet station had the ability to be a digipeater, a digital relay station that could retransmit whatever packets it heard, thus extending the range of communication. In order to use these digipeaters, however, the user had to know the callsign and/or alias of each digipeater from the beginning to the end of the route, and also had to know what order the digipeaters had to be used.

As we look at figure 1, we see how a user might connect to a distant station using these digipeaters. This system, although it was used successfully, had many limitations. In order for this to work, each station in the link had to be able to communicate directly with its neighboring station, and since the AX.25 protocol places a limit of 8 relay stations, this limits the potential range. In addition, each station in the link must be turned on and operating – this could be hit and miss since some people tend to turn their systems off when they aren't operating.

Another defect with this system is that once the link is established through these relay stations, failure of any one station would effectively cause the complete failure of the link. This failure could be due to natural causes (lightning strike, power failure) or through human causes (the user turned off his TNC or changed frequencies on his radio).

Although figure 1 shows that there may be more than one path from point A to point B, the packet system has no means to select the "best" or most reliable path, nor is it capable of detecting failures and routing the data around the failure. With the release by Software 2000 of the NET/ROM firmware, a new era of packet radio began.

FIGURE 1

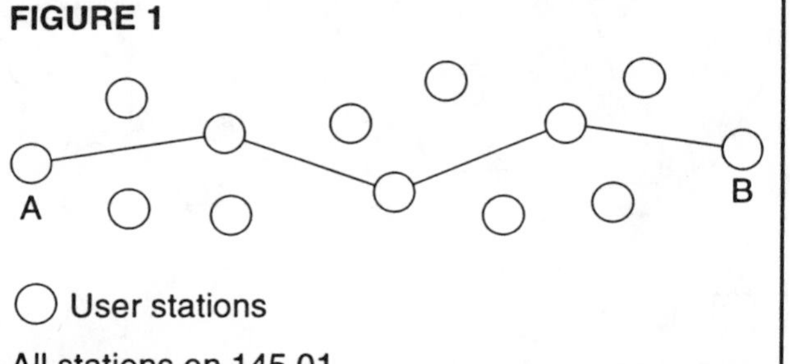

A

B

User stations

All stations on 145.01

The Beginning of Packet Networks

By replacing the EPROM firmware in the relay stations, these stations became intelligent for routing data from point A to point B. The new firmware converts these TNCs from simple digipeaters into Network "nodes". The nodes broadcast a simple packet on a periodic basis, typically once each hour, and thus other nodes recognize the presence of neighboring nodes automatically. Each node maintains a table of all of the other nodes it can hear (its neighbor nodes), and broadcasts a list of these nodes. Each node would listen to the neighboring node broadcasts, and thus learn of distant nodes – nodes which cannot be heard directly, but which the neighbor can hear. For instance, if Node A can talk to Node B, and Node B can talk to Node C, then Node A can talk to Node C (by relaying through Node B).

The initial network was built using existing 1200 baud TNCs connected to 2-meter radios, and operating on the same frequency that packet operators had been using for some 5 years – 145.01 MHz. This provided a relatively inexpensive means to improve the packet system – no new investment in TNCs, radios, or other hardware was required – only a new EPROM. As we look at figure 2, we see how the network may have looked in the beginning. All stations were operating on a single frequency, and all were operating at 1200 baud.

Problems quickly arose with this system. Since all nodes and all users were on the same frequency, the frequency quickly became overloaded with data. Remember, each node broadcasts its list of other known nodes periodically and the users were still trying to use the same frequency. Additionally, many nodes were installed at relatively high altitudes enabling them to talk greater distances, but the additional height added to the hidden station problem. A user attempting to connect to a node may suffer collisions from a neighboring node many miles away.

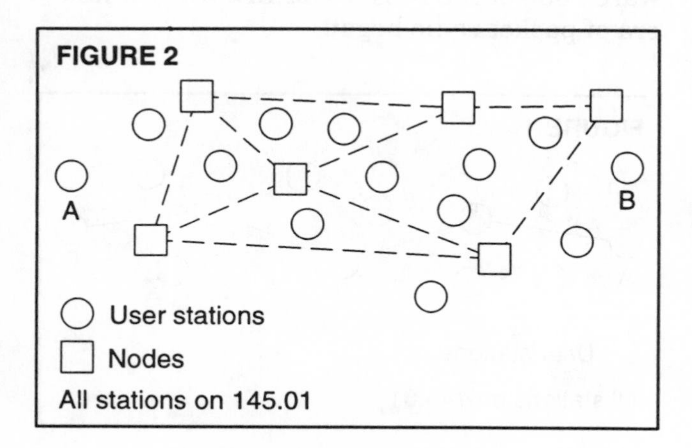

A Better Way

To overcome some of these problems, node operators began building node "stacks" – two or more nodes at the same physical location connected to each other through the serial port of the TNCs. These node stacks would have one node dedicated to "user" access, and the second node, on a different frequency, for node-to-node or "backbone" communication. This scheme allowed the data to pass with little contention over the backbone, and also reduced the hidden station problem since local users were no longer sharing the frequency with distant nodes. As more users and more nodes began to appear, node operators extended the node stacks with additional TNCs. The additional nodes were on different frequencies or bands, and in some cases operating at higher speeds.

High speed backbone systems are seen as one of the keys to success of a network system. Unfortunately, at that time, 9600 baud packet radio was in its infancy, and there were no commercially available radios capable of operating these speeds. Node operators, in their quest for higher speeds, modified existing radios to permit the higher speeds. Today, 9600 baud is rapidly being introduced into networks – partly because the current system is severely overloaded, and partly because the major radio manufacturers now produce radios capable of operating 9600 baud without modification.

One Network Plan

Figure 3 shows one possible plan for developing a network. Each node in the system consists of at least 2 ports – one for the Local Area Network (LAN) to provide user access, and a second for the backbone. The user access port is a 1200 baud port, while the backbone port typically runs at higher speeds –

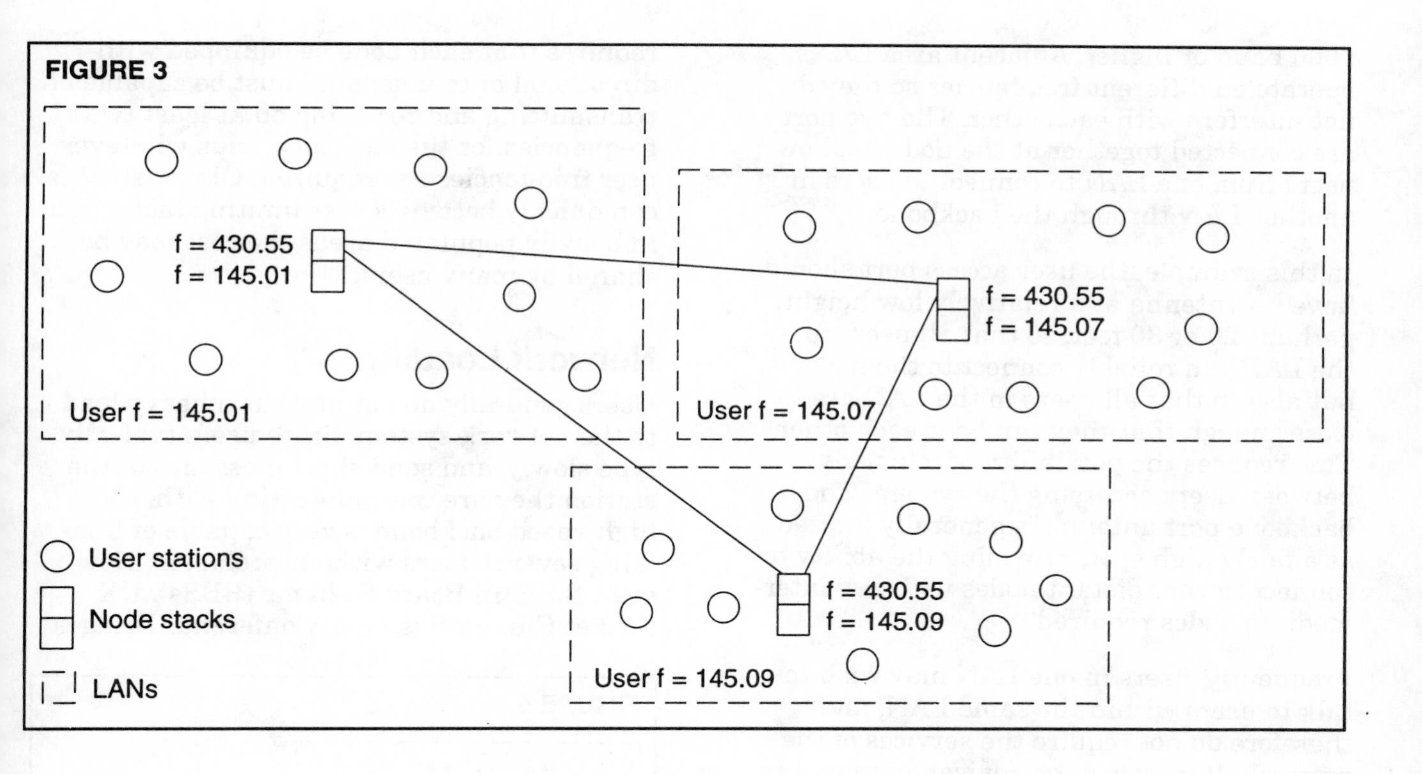
FIGURE 3
f = 430.55
f = 145.01
f = 430.55
f = 145.07
User f = 145.01
User f = 145.07
User stations
Node stacks
LANs
f = 430.55
f = 145.09
User f = 145.09

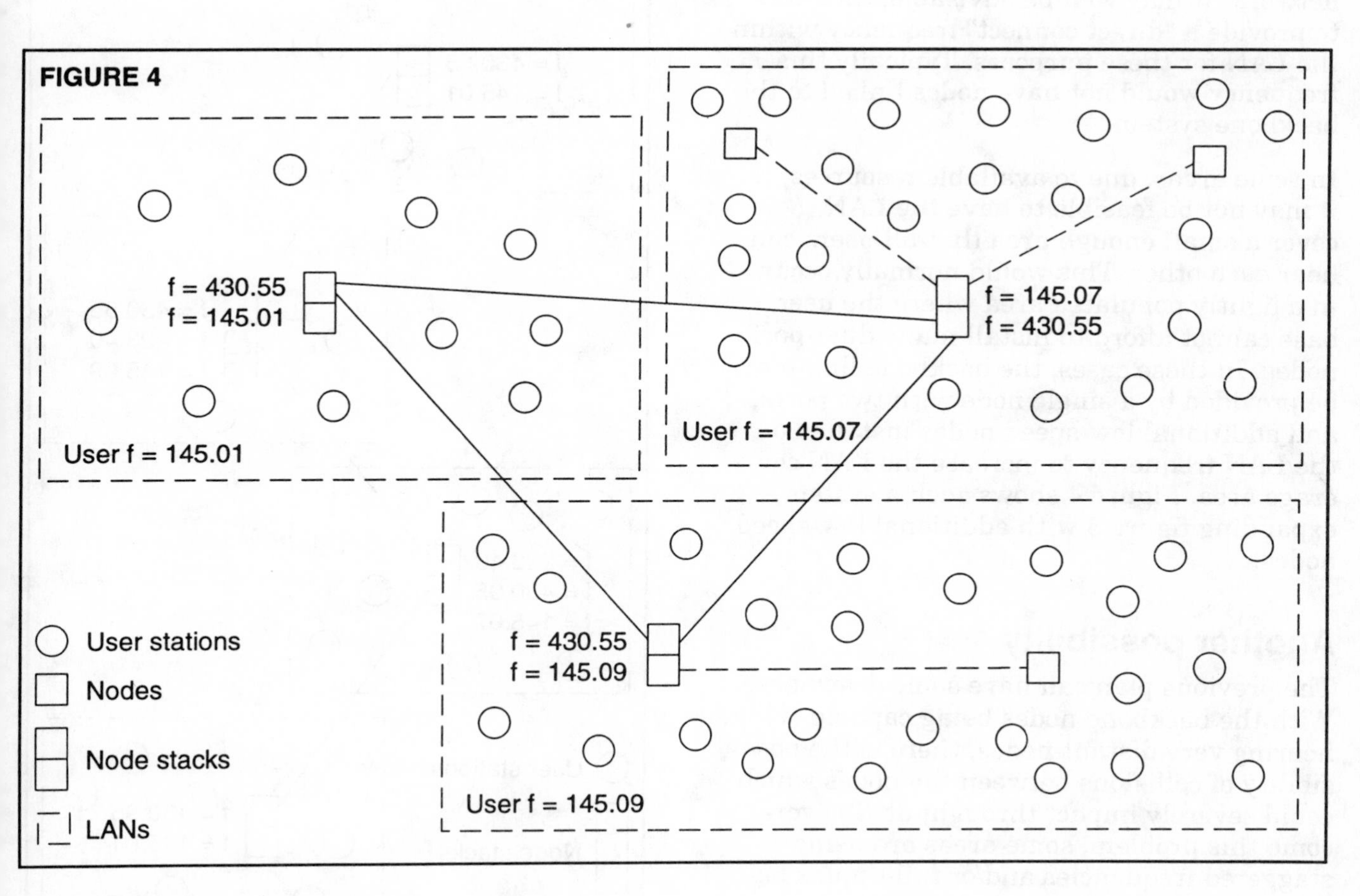
FIGURE 4
f = 430.55
f = 145.01
f = 145.07
f = 430.55
User f = 145.01
User f = 145.07
User stations
Nodes
Node stacks
LANs
f = 430.55
f = 145.09
User f = 145.09

9600 baud or higher. Adjacent area LANs operate on different frequencies so they do not interfere with each other. The two ports are connected together at the node to allow users from one LAN to connect to users in another LAN through the backbone.

In this example, the user access port should have its antenna at a relatively low height, perhaps 20 or 30 feet, so that all users in the LAN can reliably connect to the node, but also so that all users in the LAN are close enough that they can hear each other. This reduces the possibility of collisions between users accessing the system. The backbone port antenna is generally located at a fairly high spot, providing the ability to connect to very distant nodes with few intermediate nodes required.

Frequently, users in one LAN may wish to talk to users within the same LAN, and therefore do not require the services of the network. It may well be advisable, therefore, to provide a "direct connect" frequency within the LAN for these purposes. Typically this frequency would not have nodes linked to the backbone system.

In some areas, due to available resources, it may not be feasible to have the LAN cover a small enough area that all users can hear each other. This would normally occur in a lightly populated area where the user base cannot afford to install many dual-port nodes. In these cases, the backbone link may be provided by a single node with two ports, and additional low-speed nodes installed on the LAN frequency to increase the LAN coverage area. Figure 4 shows such a system, expanding figure 3 with additional low-speed nodes.

Another possibility

The previous plan can have some drawbacks. With the backbone nodes being capable of hearing very distant nodes, there is the possibility of collisions between the nodes which could severely impact throughput. To overcome this problem, some areas are using staggered frequencies and/or full-duplex between the nodes. Figure 5 shows a possible configuration using this scheme. This scheme requires that each node be equipped with directional antennas, and must be capable of transmitting and receiving on at least two frequencies for the backbone, plus whatever user frequencies are required. Obviously this can quickly become a cost-limiting factor, but in heavily populated areas the cost may be shared by many users.

Network Loading

Users generally do not present a heavy load to the network system. Since users typically type slowly, and send short messages to the station they are communicating with, the high-speed backbone is very capable of handling several users without problems. However, Bulletin Board Systems (BBSs), DX Packet Cluster Systems, Conference Bridges

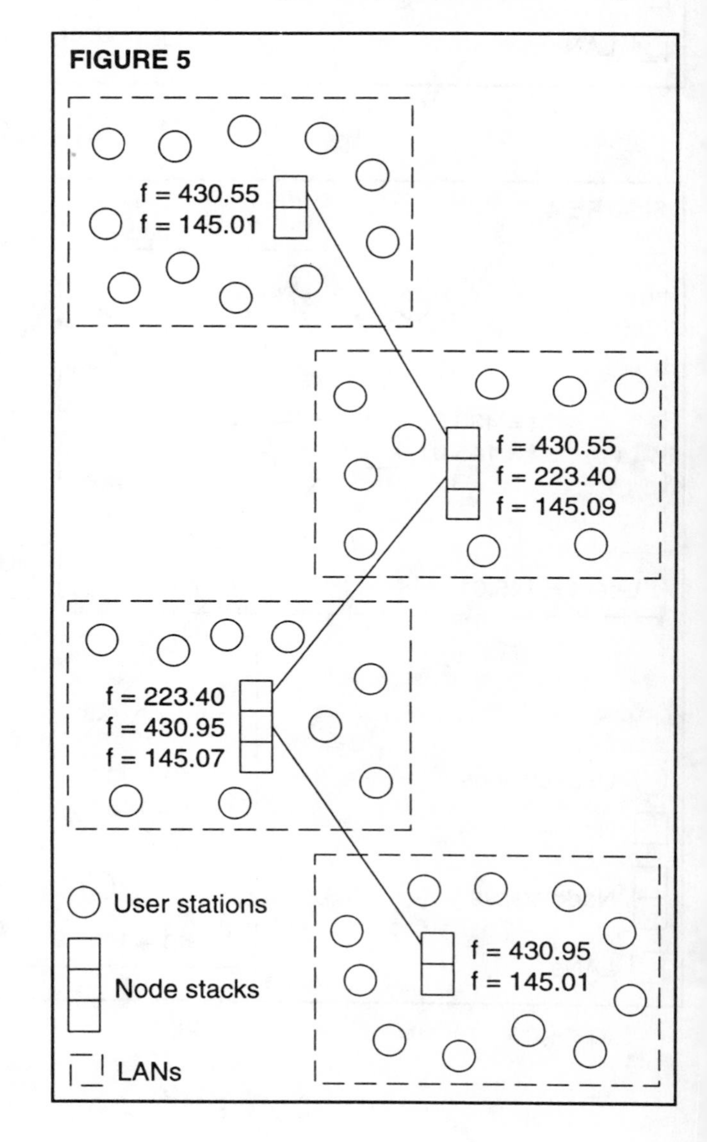

and the like present a very different situation to the network.

BBSs provide users with a tremendous amount of information – regardless of what you may think of the value of that information. This data arrives at the BBS from another BBS and so on, which means that the data has to travel through the network at some point. Within a LAN, users access the BBS to retrieve information, but there may be only one BBS in that LAN. Therefore the BBS must use the backbone to send data to other BBSs. This presents a couple of questions to consider: 1) should users access the BBS on the LAN frequency, or should a separate frequency be assigned for user access to the BBS services, and 2) should the BBS forward traffic through the network via the low speed (user) access port of the node, or should the BBS have direct access to the backbone frequency.

There are no concrete answers to these questions, they must be decided on a local basis. Things to consider include the channel loading caused by users accessing the BBS, the cost of the equipment required, available frequencies in the area, and perhaps other "political" factors.

Node Ownership

When networks first started, all that was required was a simple TNC, some new firmware, one radio, and a two-meter antenna. The overall investment to set up a node was relatively small. Those who built the network already had a TNC, and the additional investment for a used radio, small antenna, and cheap power supply was no problem. However, as the network grew, node stacks started appearing, and the cost of multi-port nodes grows quickly. For example, a single two-port node might consist of

1 – 1200 baud packet TNC	$130
1 – 9600 baud packet TNC	$200
1 – 2-meter radio	$300
1 – 2-meter antenna	$50
1 – 70-cm radio	$350
1 – 70-cm antenna	$75
2 – power supplies	$150
Feed line	$250

This brings a simple two-port node to a minimum cost around $1500. Not very many individuals will spend this amount to provide a hilltop node, and if there is a need for a 3- or 4-port node, the cost becomes prohibitive. If you are lucky enough to find a person willing to provide such a node for your system, you're very fortunate, indeed. Consider, however, that private ownership of the nodes comes with potential problems: what if the owner dies, or gets tired of packet, or needs the cash from selling the gear, or moves? You may suddenly find a hole in your network that you can't easily replace.

Perhaps a better solution is for packet groups to sponsor nodes within the area. In such cases, the packet group owns and operates the nodes, so one member of the group moving or tiring of the mode doesn't disrupt the network.

Network Parameters

So now that you've built a network, how does it automatically select the best path from point A to point B? The answer lies in the parameter settings within each node. Network nodes listen for the node broadcast from all neighboring nodes, assign a "quality" to each neighbor, and calculate a quality to each destination the neighbor has reported. These quality figures are then used in the routing of packet data to its final destination.

There are probably as many theories about setting network parameters as there are network node operators. This paper describes one such idea that is being used in the Kansas network and seems to provide reliable node operation.

Quality figures can range from 0 to 255, with 255 being "perfect" and 0 being totally unusable. As we set the quality to a neighbor, there are several items to consider. First, what speed is the link to this neighbor. A 9600 baud link is better than 1200 baud, but 56 KB is even better. If the backbone (link) frequency is on 2-meters and users are also allowed on this frequency, this is not as good as a 2-meter link that is closed to users. Perhaps the link is on 70-cm, which typically has fewer users and therefore should be a higher quality. What is the distance between nodes, and how reliable are the connections between these stations (don't forget to account for varying weather conditions here). A link may be 100% reliable during the winter months when there are no leaves on the trees, and fail miserably when the leaves come out. Rain and high humidity can also affect the reliability of a link.

Given the above, we developed the following guidelines (this is only a portion of the guidelines):

1. A 70-cm link with high-reliability, no users, and operating at 19,200 baud will be assigned a quality of 220.
2. A 70-cm link with high-reliability, no users and operating at 9600 baud is assigned a 200 quality.
3. A 2-meter link with high-reliability, users allowed and operating at 1200 baud is assigned a 140 quality.

These high-reliability neighbors are "locked in" at the quality shown, and the node is configured so that other nodes heard on these frequencies are assigned 1/2 the quality of the locked in nodes. Figure 6 shows this scheme and the resulting quality for each node.

Using this scheme, during band openings, the distant nodes that are heard are assigned a relatively low quality, causing the node to continue to use a more reliable path whenever possible. Although this will tend to limit the number of nodes in the nodes list, a user attempting to connect to one of the distant nodes will probably succeed.

Building A Network

Since the release of NET/ROM in the late 1980's, there have been many derivatives of the networking protocol. The first of these was TheNet, introduced by NORD><LINK. Other derivatives that followed were the G8BPQ code, developed by John Wiseman, TheNet Plus, various flavors of TheNet X1-J, most TCP/IP NOS programs, and now Kantronics has introduced K-Net™. All of these systems use the same networking protocol, so your network may consist of a mixture of the various types of nodes. Some of these may offer features not found in others, but the basic networking protocol remains constant.

In the past, installing a node meant that you had to remove the standard user firmware (EPROM) in your TNC and replace it with one that was specifically written for networking. This meant that you gave up the "private" use of your TNC and dedicated it to being a node. In addition, this required you to burn your own EPROM after modifying certain bytes in the image to contain your callsign, node alias, and your chosen parameters.

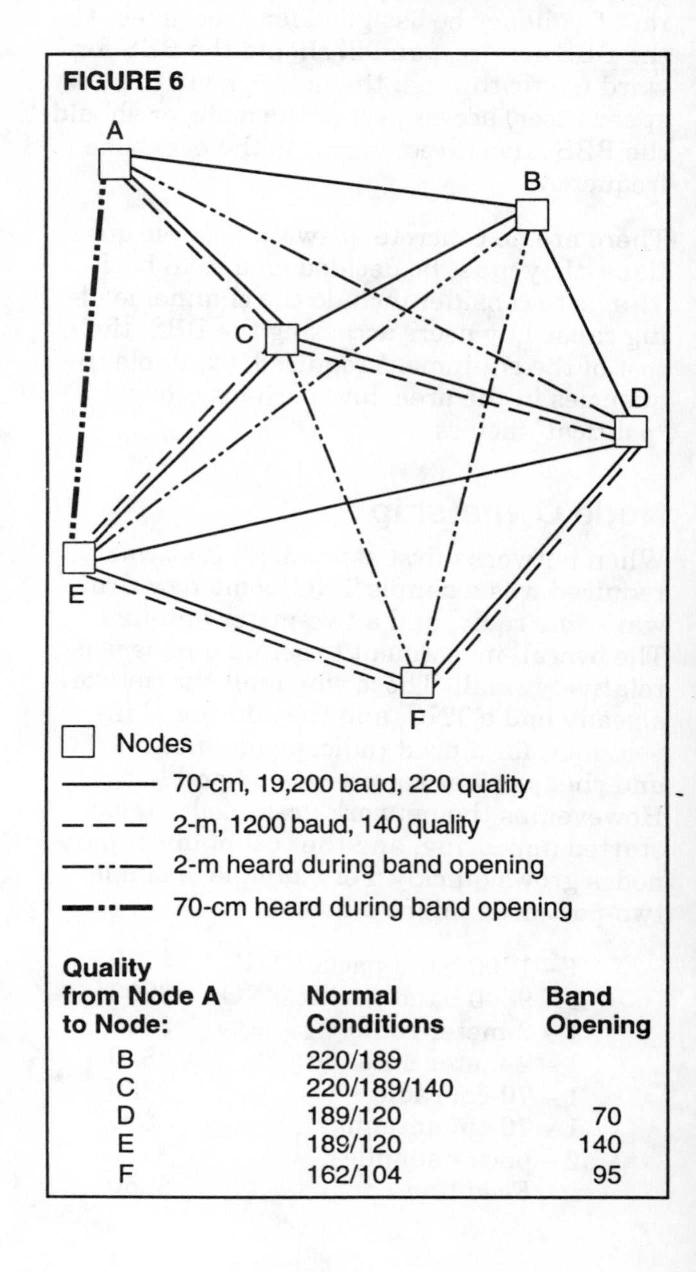

Quality from Node A to Node:	Normal Conditions	Band Opening
B	220/189	
C	220/189/140	
D	189/120	70
E	189/120	140
F	162/104	95

To create a two-port node, you needed two TNCs, each equipped with the special EPROM (but each EPROM needed its own callsign and alias). These two TNCs were connected together through their serial ports. As you expanded to a three- or four-port node, more callsigns/aliases were required, and connecting the serial ports together required a somewhat complex diode matrix to avoid data collisions on the serial ports. The number of diodes required is determined by the formula 2*N*(N-1) where N is the number of TNCs being connected together. For a three-port node, that's 12 diodes and for a four-port node you need 24 diodes.

With the release of Kantronics K-Net for the KPC-9612, users now have the ability to build a two-port node with only one TNC, providing both a 1200 baud user-access port and a 9600 (or even 19,200) baud backbone port. Since both ports reside in the KPC-9612, the node requires only one callsign and one alias. The unique features of the K-Net in the KPC-9612 don't stop here though – the TNC can still function as an end-user TNC. The PBBS in the KPC-9612 is still usable (there's a BBS command on the K-Net node) and a user can still run multiple connects, a KA-Node™, and even operate Host mode with specialized software.

Configuring a K-Net node is unlike other nodes – all the necessary commands are available at the command prompt. There is no need to patch the EPROM with the callsign and alias as other nodes require. Couple this with the remote access capability and you can place the K-Net/KPC-9612 on a remote hilltop. When the current trustee moves, an authorized user can connect remotely and change the callsign and alias of the node without making a trip to the site. In addition, the K-Net supports the NET/ROM interface over the RS-232 serial port, so it can be easily connected to an existing node stack using other brands of TNCs.

Building a four-port node with two 1200 baud ports and two 9600 baud ports would simply require two KPC-9612s with the K-Net firmware, and an interconnecting RS-232 cable with only three wires – ground, TXD, and RXD (TXD and RXD must be crossed). If you only require one high-speed port but need two 1200 baud user ports, that can be accomplished using a KPC-9612 and a KPC-3 with the optional K-Net firmware EPROMS.

Building or expanding network systems is not extremely difficult, but for the network to operate smoothly and provide maximum benefit, it requires cooperation and coordination between the node operators, BBS system operators, and other network service providers.

DAMA – Another approach

There has been some discussion and previous papers presented on DAMA (Demand Assigned Multiple Access). While this sounds like a completely different networking scheme, it really isn't. The basic premise of DAMA is twofold: 1) Prevent collisions between stations using the node, and 2) Allow additional time for stations sending more data than for those sending little data.

A DAMA node (called a DAMA MASTER) is actually a LAN node with at least one link to a backbone network. A single DAMA master can support up to 16 TNCs (thus 16 radios), perhaps providing two or three backbone link frequencies, a couple of user frequencies, some frequencies for BBS systems, and so on. Each LAN has a DAMA master node, and the adjacent DAMA masters are linked together. This inter-node link uses NET/ROM protocol, exactly like the networks just discussed.

Within a LAN, users connect to the DAMA master. Through special coding in the AX.25 frame, the user's TNC is placed into a "DAMA SLAVE" mode. From this point on, the user's TNC will not transmit any packets until the master has instructed it to do so by sending a poll frame to the slave. In this manner, no two slave TNCs will transmit at the same time, eliminating collisions.

The demand access feature of DAMA comes through intelligence built in to the master node. When the master polls a user, the user must respond – even if it's only an ack saying "I have no data to send". After polling station A, the master polls station B, then C, and so on. Any station that does not have data to send has its priority reduced. The master then polls those stations with higher priority more frequently than those with low priority. As soon as a low priority station responds to the poll by sending data, the master bumps that user's priority back up to the maximum.

The DAMA system may provide an excellent alternative in areas where LANs are very large, resulting in many LAN users being unable to hear all other users. The major drawback to the DAMA system, at this point, is that the DAMA master station requires a computer at the node site to run the special software. In addition, multiple frequency operation requires one TNC for each frequency with special DAMA firmware installed.

Users accessing DAMA systems should be using special DAMA slave TNCs. The DAMA software as distributed contains EPROM images for TNC-2 clones, enabling them to be DAMA slaves. These EPROMS use a slightly modified version of the WA8DED Host mode to communicate to the computer, and some software packages are readily available for use with these. Among the more popular programs in this category are Grafik Packet and ESKAY.

Kantronics will soon be releasing a new DAMA capable EPROM for some models of their TNCs. The new DAMA EPROM will allow users access to DAMA networks as a slave using any terminal program, dumb terminal, or even Kantronics Host mode.

Conclusion

Networking in the amateur packet world can be considered to be still in its infancy. The first NET/ROM firmware was released around 1987, and many improvements have followed. Other networking systems have been developed (ROSE and TCP/IP) for amateur use, but to-date, none has been declared the clear "winner" for packet networking.

Debates rage about the "best" or "perfect" network, but if such a system already existed, world-wide VHF packet would be a reality instead of a dream.

K-Net and KA-Node are trademarks of Kantronics Co., Inc.

NET/ROM is a trademark of Software 2000, Inc.

The Puget Sound Amateur Radio TCP/IP Network

Steve Stroh N8GNJ

Puget Sound Amateur Radio TCP/IP Group
14919 NE 163rd Street
Woodinville, WA 98072 USA
e-mail: **strohs@halcyon.com**

Abstract

The Puget Sound Amateur Radio TCP/IP Network (also known as WETNET, the Washington Experimenter's Tcp/ip NETwork), centered in the Seattle, Washington metropolitan area, has built an extremely functional packet radio network based on TCP/IP networking and cellular RF techniques. The network encompasses more than eighteen separate Local Area Networks, an estimated 200 users, four 9600 baud bit regenerative repeaters, and a full time Internet gateway. This paper is intended to provide an overview of an operational Amateur Radio TCP/IP network.

A Frequently Asked Question is "Why TCP/IP? Can't you do everything with regular packet radio that you can do with TCP/IP?" The answer is... somewhat, but not really. A key feature about Amateur Radio implementations of TCP/IP is that its various capabilities are built in- you don't have to combine dissimilar systems to do mail forwarding, file transfers, provide Packet Bulletin Board System (PBBS) services, multi-connect chat sessions, multiple ports, etc.- "It's (all) in there". A typical TCP/IP station can do:

- file transfers (including binary files)
- keyboard to keyboard (chat, telnet)
- finger (display short text files)
- electronic mail
- automatic routing
- ping (test link integrity)
- operate as a very capable Packet Bulletin Board System
- operate as a very capable Net/ROM, TheNET, X1J node, etc.
- accept multiple connections from AX.25, Net/ROM, and other TCP/IP stations
- access multiple ports, including modem, RS-232, terminals, and Ethernet connections

These operations are simultaneous, since the TCP/IP software is written to multitask. The capabilities outlined above are only a subset of TCP/IP's capabilities.

Some Background on Amateur Radio TCP/IP

A complete discussion of all of the various TCP/IP utilities and capabilities is beyond the scope of this paper, but a few deserve some discussion. One of the most interesting capabilities of Amateur Radio TCP/IP is e-mail. Since each TCP/IP station can send and receive e-mail, there cannot be a "choke point" in the network for message traffic- each TCP/IP user has the ability to send e-mail to any other user from their station. This was a much appreciated feature after having been through "PBBS Forwarding Battles", where certain PBBS Sysops wouldn't forward to other PBBS Sysops. With the added capability of a mailing list- being able to forward a single message to multiple recipients, e-mail made PBBS', and their problems, almost irrelevant. TCP/IP users have the capability to participate in PBBS forwarding, and can "translate" PBBS messages to e-mail messages, and vice-versa.

Being able to add multiple ports easily is liberating. TCP/IP users can gateway between multiple ports, and this capability is used to provide redundant routes, and to provide network access to remote Local Area Networks whose users cannot access other parts of the network directly.

A TCP/IP station can effectively service AX.25 users by providing the capabilities of a PBBS, a network node, a "chat node" , and much more to AX.25 users.

Amateur Radio TCP/IP authors have traditionally (begun by Phil Karn KA9Q) made all source code, (written in C) freely available. If you want to add a new feature, or fix a bug, the source code is available, and several Puget Sound TCP/IP operators have made good use of this to add features and fix bugs, and contribute back to the code pool.

Network Operating System (NOS) was the name that KA9Q gave to his "second generation" of Amateur Radio TCP/IP software (the first was named NET). Since NOS was released, there have been numerous offshoots of NOS, including JNOS, TNOS, and others. In this paper, NOS is used interchangeably with the term "Amateur Radio TCP/IP Software".

It is important to remember that "Amateur Radio" TCP/IP is completely interoperable with "commercial", or "wired" TCP/IP. TCP/IP was enhanced to make use of Amateur Radio just as it was to make use of satellite links, other networking systems, and modems.

Networking

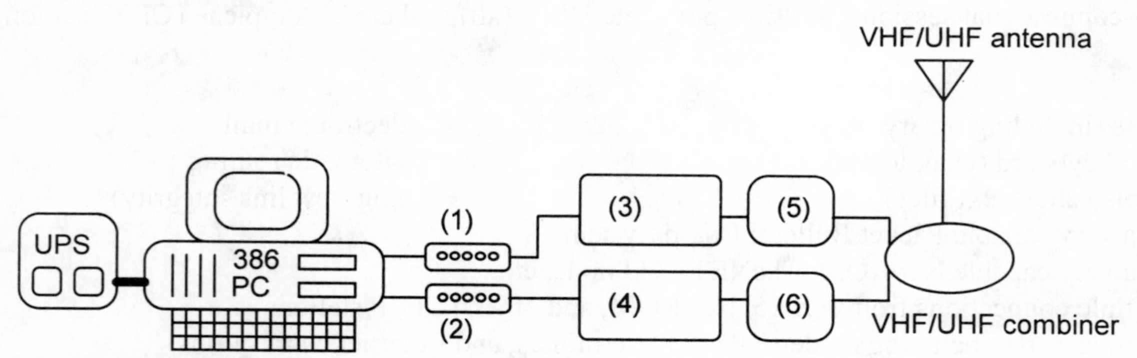

Figure 1 - Diagram of a typical Puget Sound Network Switch site

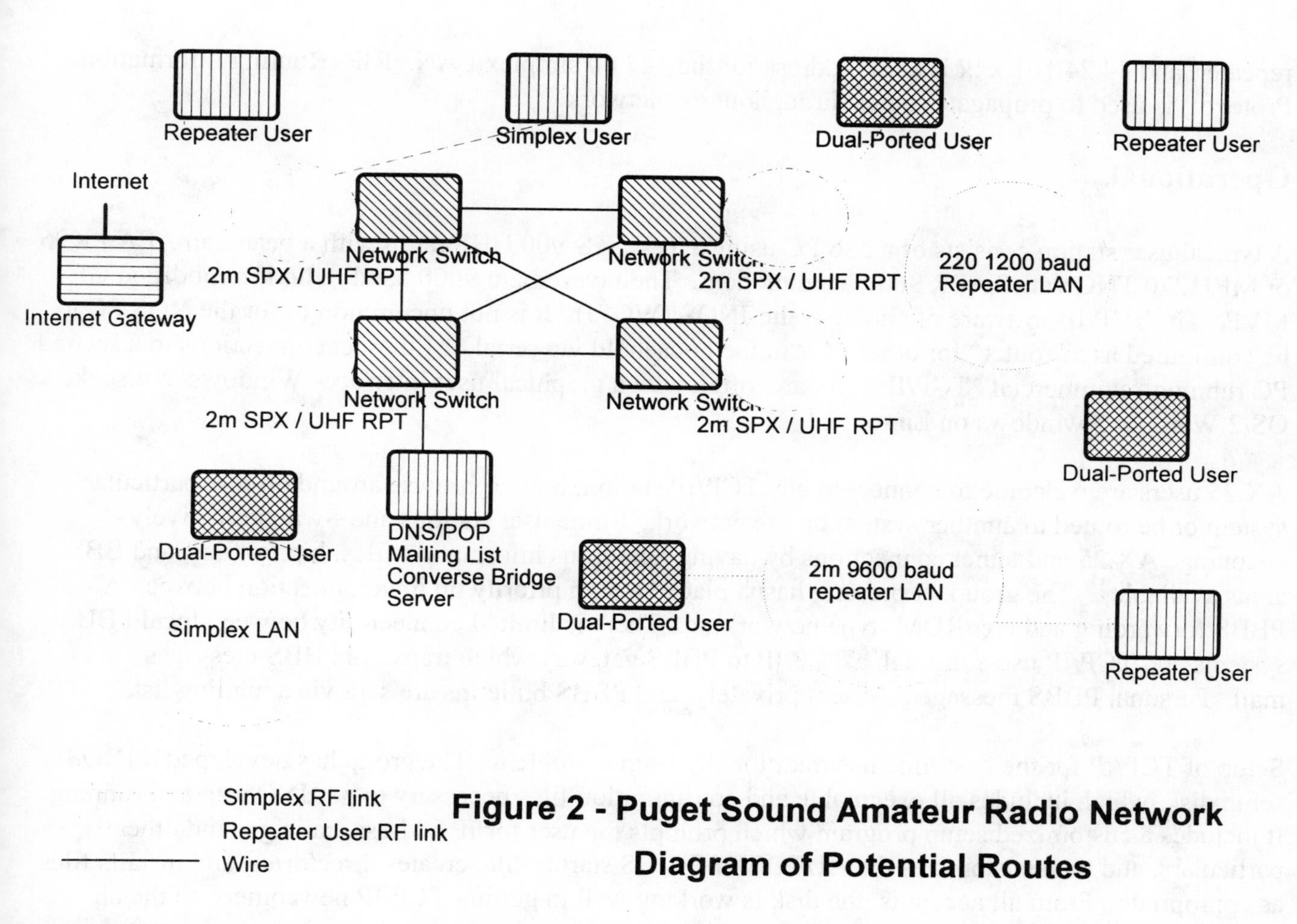

Figure 2 - Puget Sound Amateur Radio Network Diagram of Potential Routes

Typically, the Switches do not offer any "user functions" and only handle routing and other network functions. If a user function is enabled, it is typically routed to other systems in the network. This may change as the Switches are converted to Linux.

Several LANs have been formed by multi-ported stations that access a repeater and a 2m simplex frequency. Network users have gradually adopted the cellular telephone approach of using low power and a low RF profile, and thus are able to reuse frequencies. Typically, the multi-ported station forms the center of a simplex LAN. Any stations joining the LAN are encouraged to be within good DCD range of the other stations on the LAN. This keeps overall performance on the LAN reasonably high, despite the inevitable collisions and hidden transmitters resulting from simplex operation.

The network is linked full time to the Internet through a Network Switch at the premises of one of the largest Internet Service Providers in the Seattle area. The primary use of the gateway is to allow participation in the global Converse Bridges. Amateur Radio TCP/IP users can also telnet, FTP, etc. to Internet hosts, as long as the user has a good connection to the network so the Internet system doesn't time-out from long delays.

The group makes extensive use of subnetting- the third "octet" of the Amateur Radio TCP/IP IP address is specific to a particular LAN. For example, 44.24.103.xxx is an IP address for the 220 1200 baud repeater, and 44.24.101.xxx is an IP address for the 147.60 simplex LAN. RIP (Routing Information Protocol) is used to propagate routes throughout the network.

Operational

A typical user station consists of a 386 PC using a Tekk KS-900 UHF radio with a beam, an AEA PK96 or MFJ1270 TNC with TAPR 9600 baud modem. The favored 2m 9600 baud radio is a modified GE MVP. The TCP/IP software of choice is the JNOS (WG7J). It is not uncommon to for the NOS PC to be configured as a "router" for other PC's in the household via serial or Ethernet connections to a second PC running "commercial" TCP/IP software, often with a graphical user interface- Windows/Winsock, OS/2 Warp, or Xwindows on Linux.

AX.25 users are welcome to connect to any TCP/IP station, both to browse around on that particular system or be routed to another system on the network. Some user stations and Switches actively encourage AX.25 and telnet connections by having numerous information files, binary files, and BBS areas available, . The group, at present, hasn't placed a high priority on interconnection between AX.25 PBBS forwarding and Net/ROM -type networking. There is limited connectivity between local PBBS systems and TCP/IP users through a TCP/IP to PBBS gateway which transfers PBBS messages via e-mail. Personal PBBS messages are sent privately, and PBBS bulletins are sent via a mailing list.

Setup of TCP/IP for the first time has traditionally been a problem. The group has developed a JNOS setup disk, which includes all executable and configuration files necessary to get JNOS up and running. It includes a customized setup program which prompts the user for IP address, callsign, and other particulars, and then customizes the AUTOEXEC.NOS startup file, creates directories, and installs files as appropriate. From all accounts, the disk is working well in getting TCP/IP newcomers on the air.

The Puget Sound Amateur Radio TCP/IP Group

Early members of the group had been active in other area packet radio groups and had become burned out from the formal duties of a group, such as holding office, constitutions, dues, newsletters, etc. A deliberate, conscious decision was made that this group would remain informal in order to concentrate on technical and other innovative lack-of-management techniques. When funding has been needed for repairs or new equipment, money is donated by members of the group. This arrangement has worked relatively well, but the group did recently decide to get "slightly less disorganized" by requesting volunteers for *Keeper of the Notes*, *Keeper of the List of Projects*, and *Keeper of the Shekels*.

The group communicates primarily through a mailing list. Members of the group that wish to pose a question, comment on a previous posting, or share information send a single message to the list, and that message is automatically propagated. If a user's connectivity is good, messages from the list can be delivered in minutes. Mailing list functions have been automated using Majordomo mailing list server software on a Linux system.

The group meets monthly on a weeknight evening. In keeping with its informal nature, meetings are "moderated" only to the point of "collision avoidance". There is usually plenty to discuss, and formal presentations are rare. Any "burning issues" have typically been discussed on the mailing list prior to the meeting, and there are only a few "reports". The group has a small advertisement in the local "Computer Paper" which draws in a few lapsed hams and technically curious non-hams.

An ongoing effort is to write effective documentation. Those involved in the documentation effort are enthusiastic about the potential of using the World Wide Web (WWW, or Web) to distribute information. Articles can be distributed as they are created, and updated individually. Web pages, being composed of ASCII text, can easily be distributed as e-mail, or printed from a Web browser. Using Web pages also enables other groups worldwide to easily access the group's information. The group will support Web pages on the Internet gateway (typical graphically rich Web pages), and a well connected RF site in the network (primarily text Web pages). The group's current Internet Web page is located at: `http://wetnet.ampr.org/`

Future goals and projects

- Enhancing the reliability of the overall network is a high priority
- Expand high speed TCP/IP links to other areas in the Pacific Northwest, especially Vancouver and Victoria, British Columbia and Portland, Oregon.
- Implement several 56 KBPS and higher links
- Replace NOS with Linux at all Network Switches
- Increased use of Linux and other TCP/IP capable systems
- Contribute to the forthcoming TAPR TCP/IP book
- Sponsor a TAPR Annual Meeting and/or an ARRL Digital Communication Conference
- Accommodate increased use of PC's, Macs, Amigas, and Unix systems running their native implementations of TCP/IP
- Aggressively test and implement new radio-based routing techniques as they are developed

Conclusion

TCP/IP works very well for Amateur Radio networking. It has been a satisfying learning experience to get it running and keep it functioning. The results are well worth the effort expended.

The group would enjoy hearing from Amateur Radio TCP/IP users around the country and the globe that have implemented Internet access to their Amateur Radio Networks.

For those that are able, wouldn't it be more interesting to use our "radio" computers to participate in mailing list discussions such as NOSBBS, NETSIG, and others? There is no reason not to if your system has adequate connectivity to the Internet.

References

Steve Stroh N8GNJ, **strohs@halcyon.com** for questions on the Puget Sound Amateur Radio TCP/IP Group,

Ken Koster N7IPB, **n7ipb@wetnet.wa.com** for questions on the Puget Sound Amateur Radio TCP/IP Network,

Jeremy Donimirski WA7YGB, **jdonimi@uswnvg.com** for questions on the US West NewVector Amateur Radio Group and the Cellular Base Station systems.

The most complete book on Amateur Radio TCP/IP is NOSintro, subtitled TCP/IP of Packet Radio, An introduction to the KA9Q Network Operating System by Ian Wade G3NRW, copyright 1992 by Dowermain Ltd., ISBN 1-897649-00-2. NOSintro is available from the ARRL.

A good FTP site for downloading Amateur Radio TCP/IP software from the Internet is:
`ftp.ucsd.edu/hamradio/packet/tcpip`

The mailing list for Puget Sound Amateur Radio TCP/IP Network users is:
`seatcp@wetnet.ampr.org`

My thanks to the users and builders of the Puget Sound Amateur Radio TCP/IP Network- it's not a network without users to use it. It's been quite a ride, and we're "finally having fun yet". I'd especially like to thank N7IPB, KD7NM, WA7QFR, N7NKJ, and WA7FUS for inviting me into the group.

Chapter 4
Construction

An Easy Path to Packet: the IMP

A single-chip modem makes for an inexpensive packet package.

By Tony Marchese, N2YMW
35 Shannon Crescent
Spencerport, NY 14559-9758

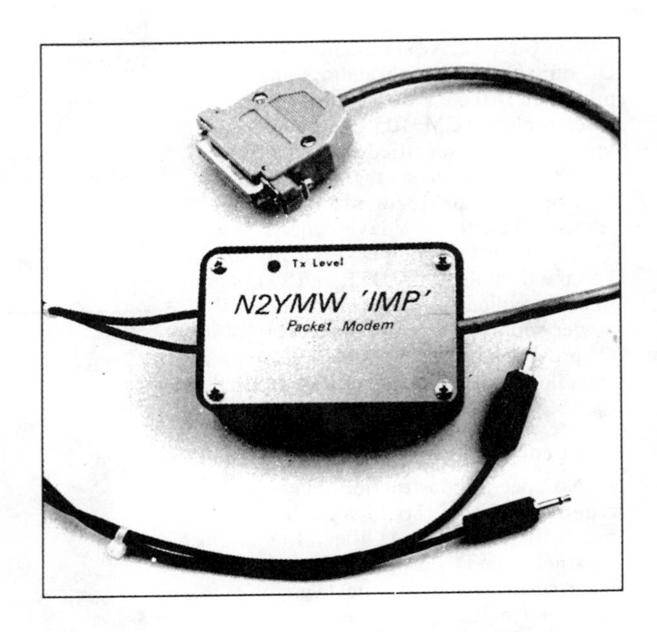

Here's a Bell 202-compatible, FSK modem I've dubbed the IMP—an Introductory Modem for Packet. The IMP can be used with any computer equipped with an RS-232-C serial port, appropriate communications software and a transceiver to access 1200-baud packet applications such as PBBSs and satellites. The IMP connects directly to the computer's serial port and typically requires no external power supply (more about that later).

Circuit Operation

Refer to Figure 1. U1 performs all of the filtering, timing, and conversion functions required for modulation and demodulation of the digital and analog signals. Q3 inverts the CLK signal to U1's transmit/receive standard select (TRS) input, which, in conjunction with bit-rate-select inputs (TXR1) and (TXR2) at ground potential, configure the TCM3105 as a 1200-baud, Bell 202-compatible modem. This configuration uses identical tones for the transmit and receive signals, with 1200 Hz representing a *mark* (logic 1) and 2200 Hz representing a *space* (logic 0), the appropriate frequencies for 1200 baud, half-duplex packet operation. Amateur packet uses a half-duplex communication mode, (ie, one direction at a time) because radio transceivers don't typically

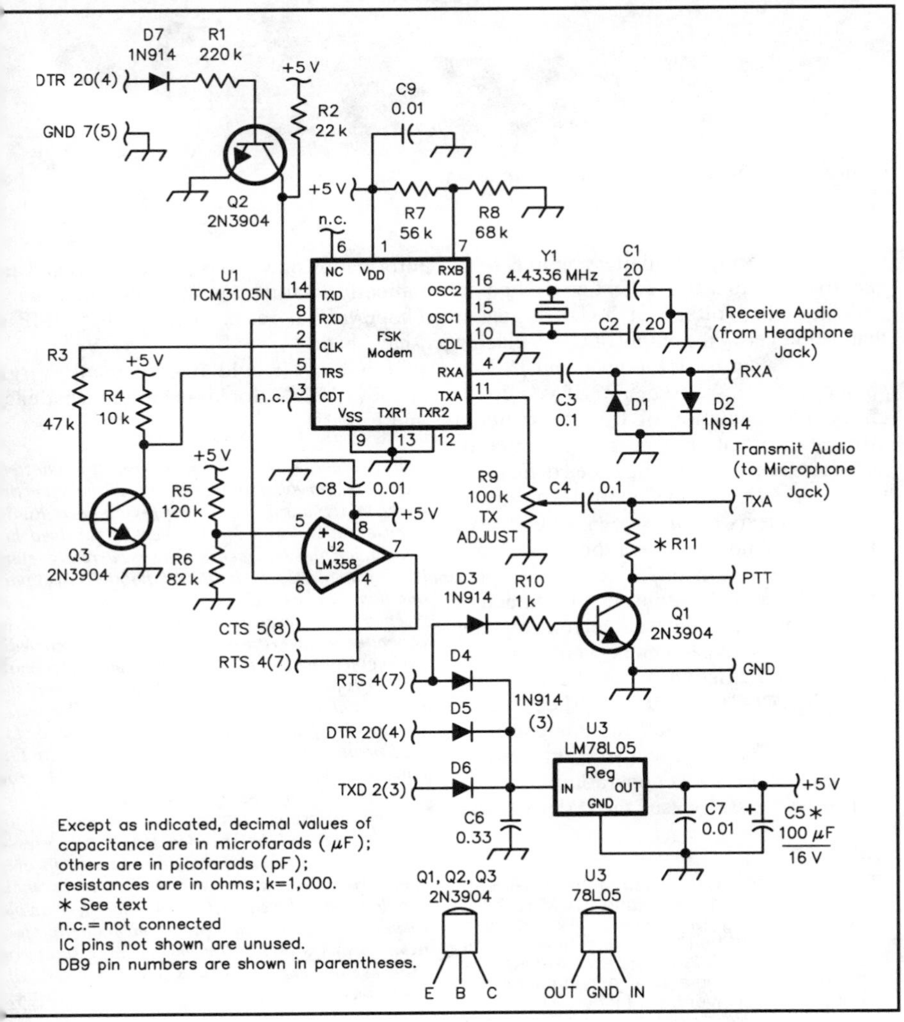

Except as indicated, decimal values of capacitance are in microfarads (μF); others are in picofarads (pF); resistances are in ohms; k=1,000.
∗ See text
n.c.= not connected
IC pins not shown are unused.
DB9 pin numbers are shown in parentheses.

Figure 1—Schematic of the IMP circuit. RS part numbers in parentheses are Radio Shack; DK part numxbers are Digi-Key (Digi-Key Corp, Box 677, Thief River Falls, MN 56701-0677; tel 800-344-4539, 218-681-6674; fax 218-681-3880) equivalent parts can be substituted. Unless otherwise specified, resistors are 1/4-W, 5%-tolerance carbon-composition or film units.

C6—100 μF, 16 V (RS 272-1029; DK P6620)
D1-D7—1N914, 1N4148 (RS 276-1122; DK1N4148CT)
Q1, Q2, Q3—2N3904A (MPS3904; DK 2N3904)
R9—100-kΩ trimmer potentiometer (RS 271-284; DK 3296W-104 or DK 3296Y-104)
R11—User-selectable for radios that combine PTT and TXA lines; refer to your transceiver's user manual.
U1—TCM3105 FSK modem (available from JDR Microdevices, 1224 S 10th St, San Jose, CA 95112, tel 800-538-5000, 408-494-1400; fax: 408-494-1420)
U2—LM358, dual, low-power comparator (DK LM358AN)
U3—LM78L05 positive 5-V, 100-mA voltage regulator (DK AN78L05)
Y1—4.4336 MHz (DK X083 [HC-49/UA holder]; DK SE3406 [cylindrical holder])
Misc: enclosure 3¼×2⅛×1⅛ (270-230); connectors for transceiver; DB25 or DB9 connector as needed; 8-pin IC socket; 16-pin IC socket

transmit and receive simultaneously. D1 and D2 limit the incoming audio signal to approximately 0.65 V as the maximum signal level for the TCM3105's Receive Analog input (RXA) is specified as 0.78 V.

Q2 and U2 convert the information signals between the ±3 to ±15 V, RS-232-C levels and the 0 to 5 V level required by the TCM3105. D4 through D6 logically **OR** the voltage from the RTS, DTR and TxD signal lines to typically produce a 7 to 15-V dc power supply. The combination of Q4 and C5 provide a relatively stable 5-V dc source capable of providing the 10 mA required by the IMP.

Construction

No special construction techniques are required and a PC board is available.[1] The board fits neatly inside a small (3¼×2⅛×1⅛-inch [HWD]) project case specified in the parts list. I built my latest version of this modem (including homebrewing the PC board and lettering the enclosure) in two evenings. You'll need a cable to connect the modem to your transceiver. Refer to your transceiver's documentation to determine the required connector types and wiring.

Adjustments

The computer requires packet communications software. The IMP is compatible with a variety of TNC-less packet software such as *BayCom,* written by Florian Radlherr, DL8MBT. *BayCom* is available from the ARRL BBS and FTP site,[2] and from many shareware distributors.

Connect your transceiver and a computer equipped with an RS-232-C port to the IMP. Turn on the radio and tune the receiver to an active packet frequency. Start the packet software, *but don't attempt to transmit.* While running the software in the *receive* mode, place a voltmeter across the modem's TxD and GND pads and verify that the voltage supplied by the computer's serial port is a minimum of +7.0 V. If the voltage is below 7.0 V, you'll need to power the IMP from an external 9-V battery or other power source.

The IMP is easily modified for 9-V battery operation. A 9-V battery connector, an SPST switch and some hook-up wire are all you'll need. Solder the positive battery wire to one side of the switch. Remove the serial port's TxD wire from the PC board. Solder a length of wire to the switch operating arm and to the TxD pad. Complete the modification by soldering the battery connector's negative lead to a convenient ground point.

When the voltage at TxD is greater than 7.0 V, verify the output of U3 by placing a voltmeter across U1 pins 1 and 9. You should measure 4.5 to 5.5 V. (If you don't, check the orientation of U3.) You should also see information appear on the monitor as data is received. You may need to increase the receiver's volume control slightly if nothing appears on the screen. The receive mode should now be operational.

The last adjustment optimizes the modem's analog output level. Attach a dummy antenna to the transceiver. Activate the software's transmit mode while monitoring the radio's output on another receiver. Adjust R9 (**TX ADJUST**) until you hear a clean signal.[3] That's it! Install the modem in its case, replace your transceiver's antenna and you're ready for packet operation!

I've successfully used the IMP to access information on various packet BBSs and am patiently awaiting my first satellite contact. I hope you have as much fun with the IMP as I have had.

Finally, I would like to thank Virgil Ansley, KF2XT, for his inspiration, patience and Elmering.

Tony Marchese, N2YMW, was born in Rochester, New York, in 1960. His interest in radio began in the early 1970s, but he didn't get his Amateur Radio license until March 1994, when he received his Technician plus HF license. Tony is currently refurbishing an old Heathkit SB300/SB400 "twins" pair and hopes to be on HF soon.

Tony holds several degrees including a BSc in computer engineering and AAS degrees in biomedical engineering and electrical engineering technology. He is presently employed as the manager of the Biomedical Engineering Department at Rochester General Hospital and as an adjunct instructor for the Computer Related Curricula and Electrical Engineering Technology Departments of Monroe Community College.

Tony has many interests, but the majority of his personal time is spent in family activities—whether it be a trip to the zoo or a home-improvement project—with his wife, Colleen, and their two daughters, 5-year-old Sarah and 1½-year-old Emma. Tony enjoys designing and building electronic projects and is presently concentrating on the development of devices for the physically challenged.

Table 1
RS-232-C Signals Used with the IMP

DB25 (DB9) Pin No.	*RS232 Signal*	*RX Level (V)*	*TX Level (V)*	*Signal from*	*Notes*
2 (3)	TxD	+10	+10	Computer	+10 V (most of the time)
4 (7)	RTS	–10	+10	Computer	Activates PTT during transmit
5 (8)	CTS	Data	X	Modem	Data from modem during receive
7 (5)	GND	0	0	Common	Signal ground
20 (4)	DTR	+10	Data	Computer	Data to modem during transmit

Figure 2—An inside view of the IMP modem.

[1] A PC board for this project is available from FAR Circuits, 18N640 Field Ct, Dundee, IL 60118-9269, tel 708-836-9148 (voice and fax). Price: $4.50 plus $1.50 shipping for up to four boards. VISA and MasterCard accepted with a $20 minimum order, or $2 service charge on orders of less than $20. A PC-board template package is available free of charge from the ARRL. Send your request for the MARCHESE IMP MODEM TEMPLATE along with a stamped business-size SASE to the Technical Department Secretary, ARRL, 225 Main St, Newington, CT 06111-1494.

[2] ARRL BBS: 860-594-0306; ARRL Internet FTP; oak.oakland.edu in the pub / hamradio / arrl / qst-binaries directory.

[3] The IMP modem was tested in the ARRL Lab using a Kenwood TH-22 and ICOM IC-2AT, the latter having a PTT switch in series with the mike line. When the mike connector was plugged into the IC-2AT, the receive mode was interrupted. Making a small adjustment to R9 brought the receiver back to life.

Radio-TNC Wiring Diagrams

The following information was graciously provided by Gloria Medcalf, KA5ZTX. She has done a fine job in compiling wiring diagrams or the most common TNCs and radios used for packet radio. To use these diagrams, wire pins with the same names to each other. For example, he Transmit Signal on the radio connector is wired to the Transmit Signal of the TNC connector. Pins are labeled with the descriptive names sed in Chapter 5 of Gloria's book, *What is Your TNC Doing?* Consult her book for information not found here. Manufacturers have been nown to change their usage of pins, so it would be wise to compare these diagrams with those given in the product manual. Damage may ccur if a wrong connection is made.

TNCs

AEA, PK-232

AEA PK-232 Radio Connector

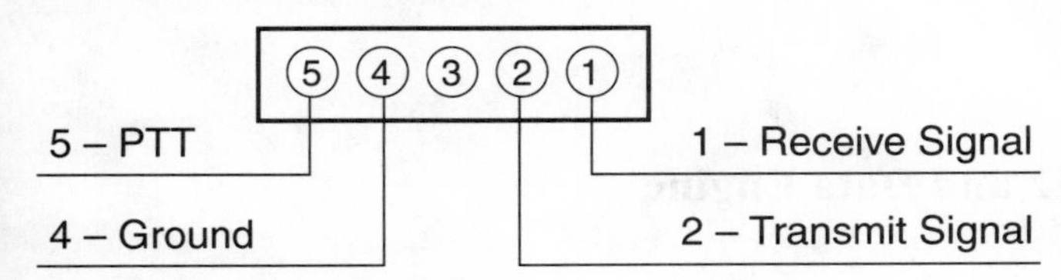

Female (wiring side)

AEA, PK-88

AEA PK-88 8-pin Mic Connector

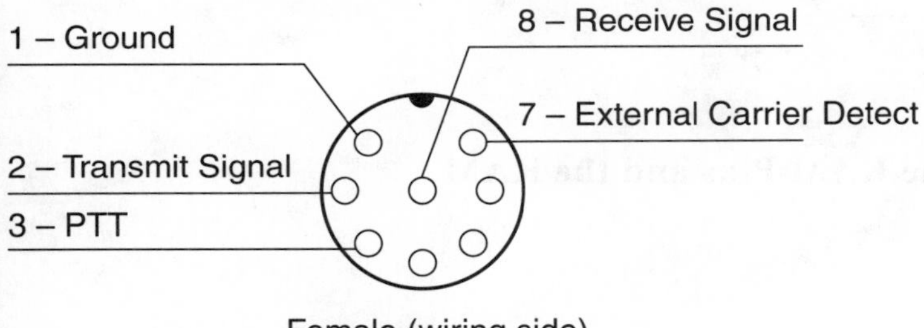

Female (wiring side)

5-pin DIN connector used on many popular TNCs including the DPK-2, DSP-2232, KPC-1, MFJ-1270, MFJ-1274, PacComm 200, PacComm 220, PK-87, PK-900, Tiny-2, TNC-2

TAPR TNC-2 5-pin DIN Connector

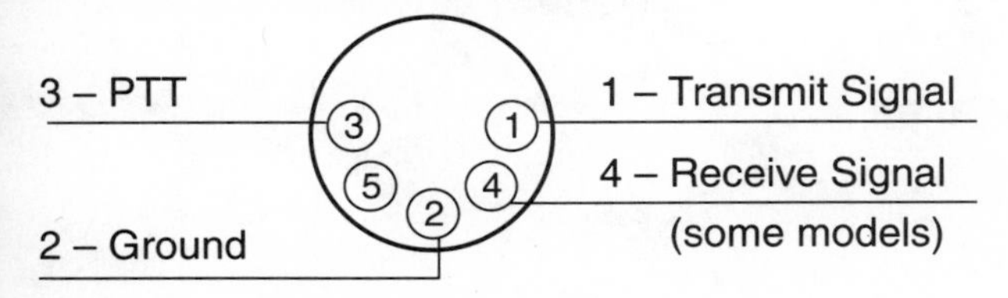

Male (wiring side)

Notes: Pin 4 on the KPC-1 has a separate connector the Received Signal (pin 4). Pin 5, External Carrier Detect, is used on the PK-900, DSP-2232, DPK-2

Kantronics with DB-9 connector for KPC-9612, KPC-3, KPC-4, KPC-2400,KAM Plus, KAM

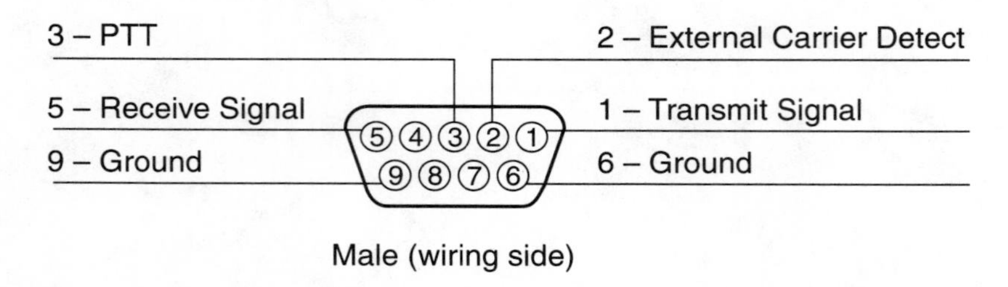

Kantronics with DB-15 connector for KPC-9612 and Data Engine

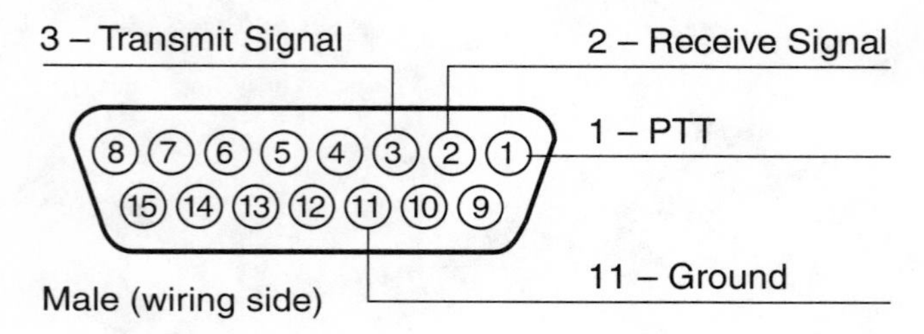

Kantronics HF 8-pin DIN connector for the KAM Plus and the KAM

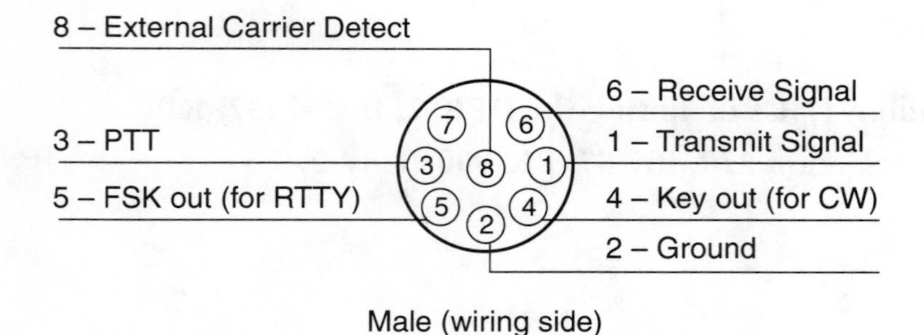

Radios

8-pin Mic Connector for Alinco and Kenwood radios

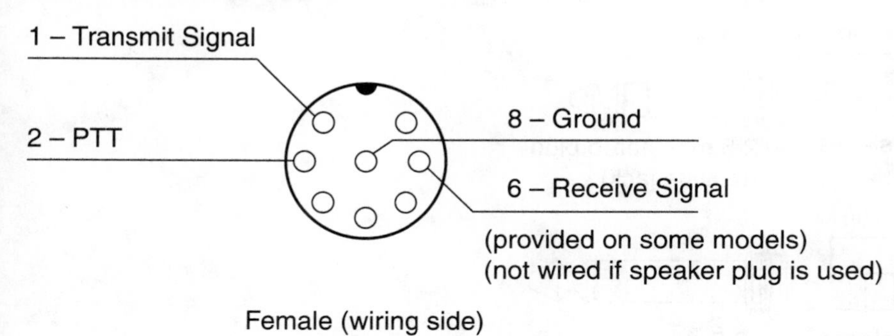

Notes: Pin 6 may not be sued on some modes that use the speaker plug.

Speaker Plug

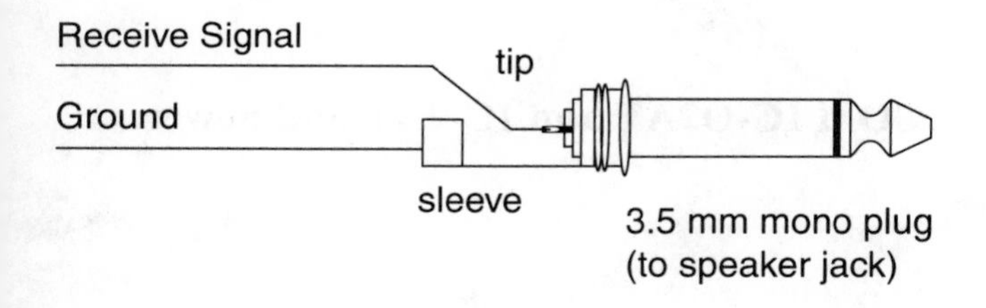

9600 Data Jack and PTT Jack for real panel connection to Alinco DR610T and DR150T

Alinco 9600 baud rear connections – DR610T

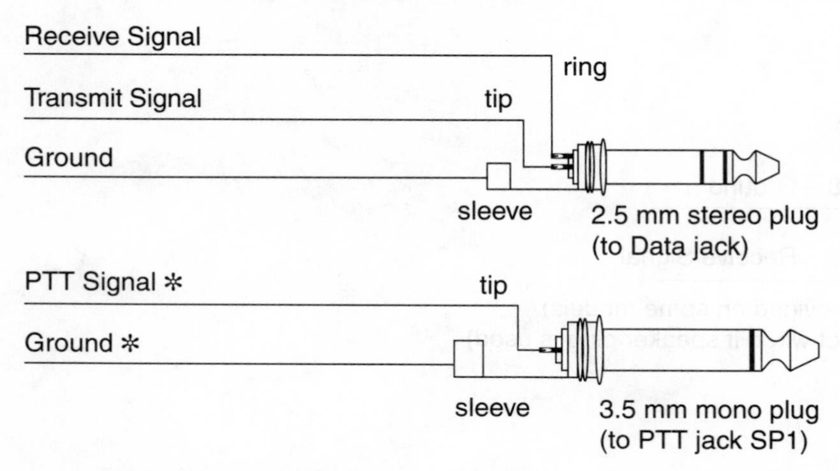

✻ Models like the DR150T use the same 2.5 plug wiring, but have PTT and Ground wired to the front-panel 8-pin mic connector.

Notes: The DR150T uses the 2.5 mm plug but PTT and Ground are wired to the front-panel 8-pin mic connector. The radio may have to be in the packet mode for the PTT to be activated on the plug. Some Alinco manuals have the PTT wiring information printed incorrectly.

Separate Mic and Speaker Plugs for Alinco and ICOM IC-O2AT and IC-2AT and newer models

ICOM 2AT Style Hand-Held Radios (and newer models)

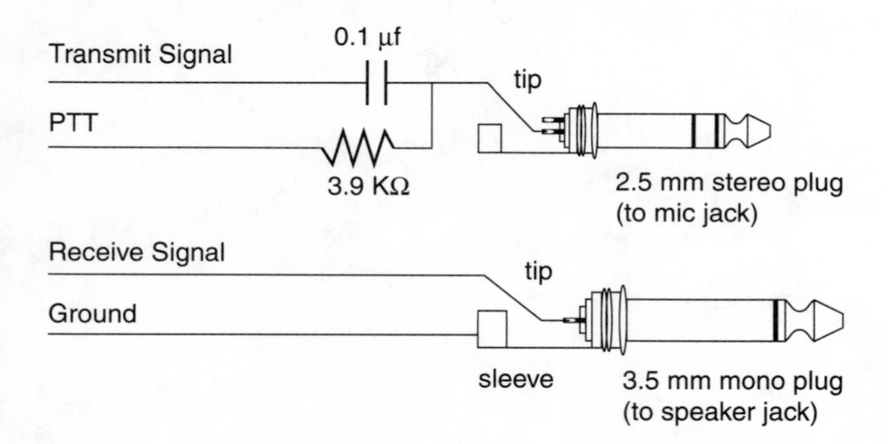

Notes: Values of the resistor and capacitor are not critical. Alinco radios may require a 15K Ω resistor instead of the 3.9 KΩ as shown.

Single Mic and Speaker Plug for Alinco and ICOM IC-W2A and newer models

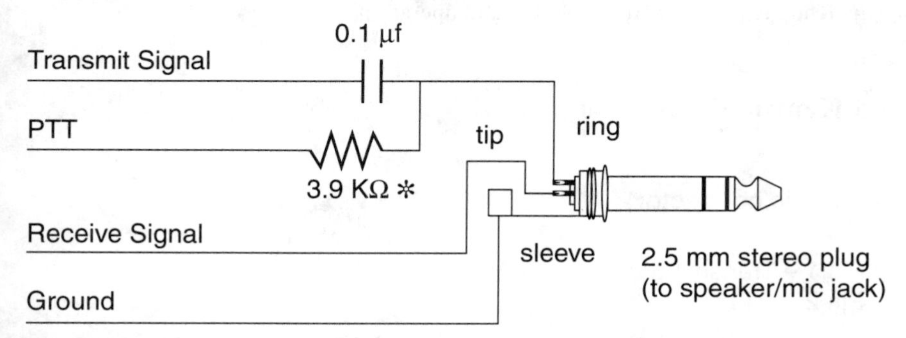

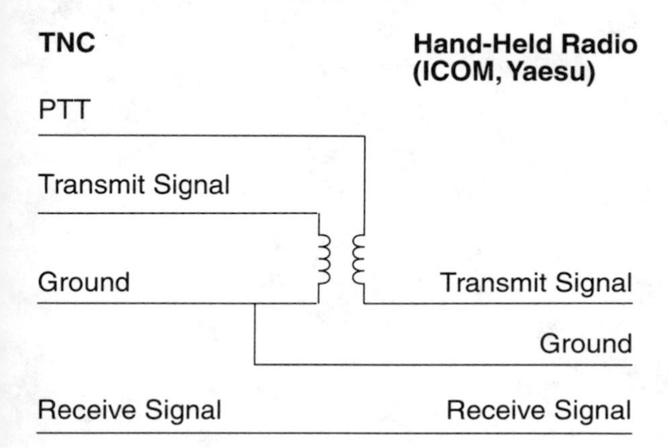

Notes: Alinco Hts may require a 15 KΩ resistor. A transformer is used instead of the resistor and capacitor for ICOM and Yaesu radios.

6-pin Mini DIN for Azden and Kenwood radios

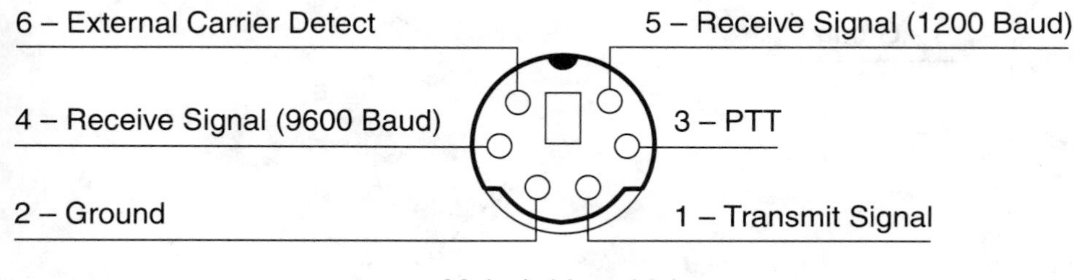

Notes: Some radios may require you to select Pin 4 or Pin 5 for the Received Signal. This is sometimes done through a menu selection. Other radios may always have both pins active so that no selection is necessary. Both speeds can be received on Pin 4, however, 1200 baud may have more noise if Pin 4 is used instead of Pin 5. The audio at Pin 5 may be unsquelched. The audio at Pin 4 is unsquelched. Many 1200 baud TNCs accept squelched audio as the default mode. If only unsquelched audio is available, the TNC will always be receiving noise and must be able to detect packet signals out of the noise. This is often accomplished with a TNC command for carrier detect mode. Unsquelched is the standard for 9600 baud operation.

Mic Schematic for Drake TR-7 and Kenwood TR-7400

Typical Schematic of Microphone (4-pin mic connector)

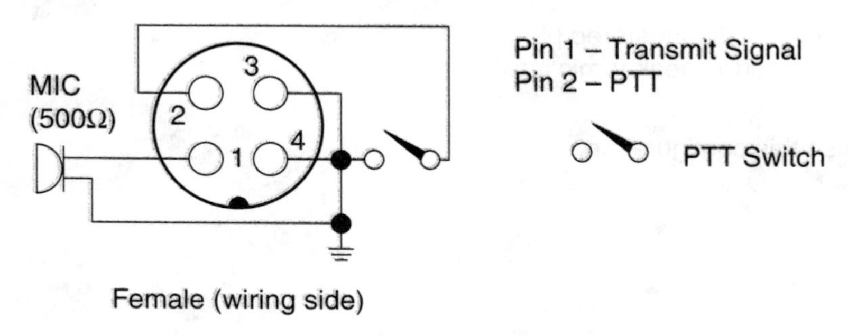

Typical Speaker Plug

Typical Connection for Receive Signal to Speaker Plug

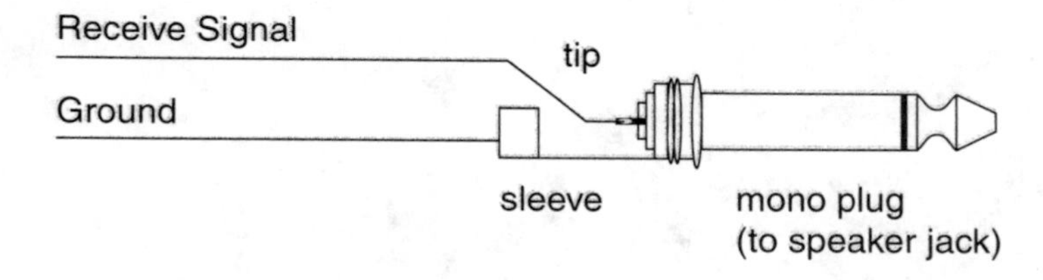

ICOM 8-pin Mic Connector

Icom 8-pin Mic Connector

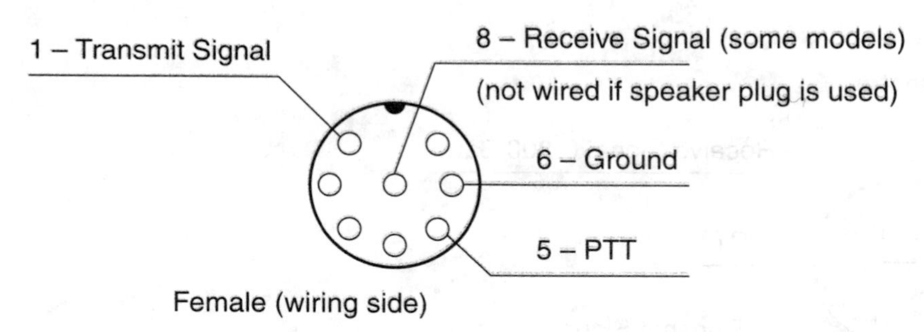

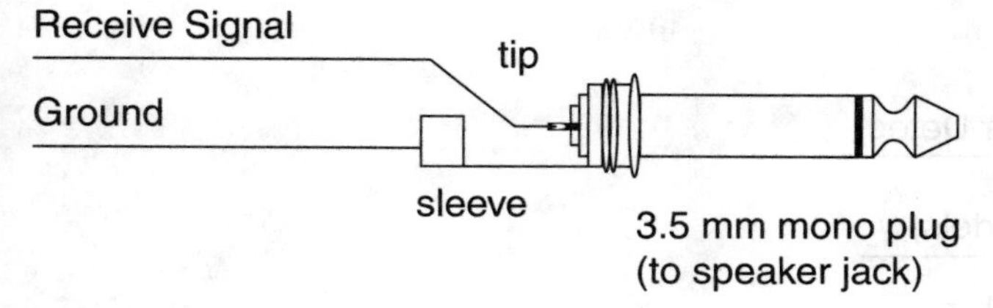

Notes: Pin 8 is not used if the external Speaker Plug is used.

ICOM 8-pin Modular (RJ-45) Mic Connector

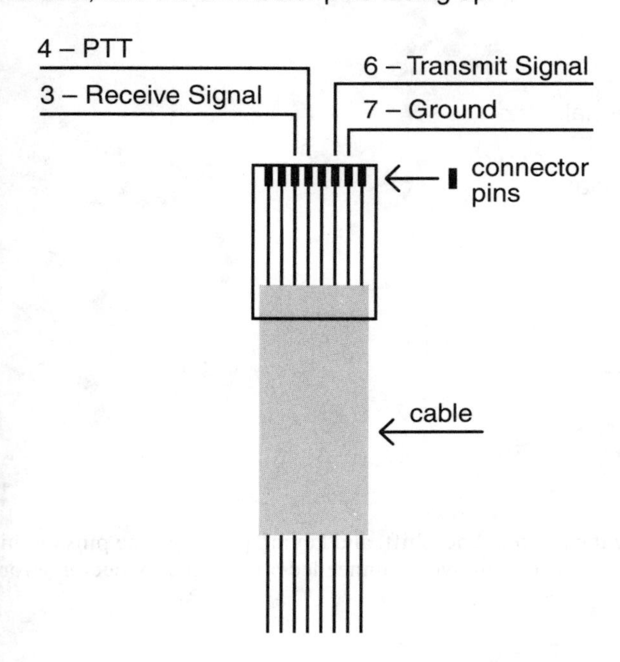

Notes: Pins are numbered 1 to 8 from left to right when looking at connector with the cable towards you, the locking mechanism down, and the connector pin facing up.

ICOM Radios with 8-pin accessory connector

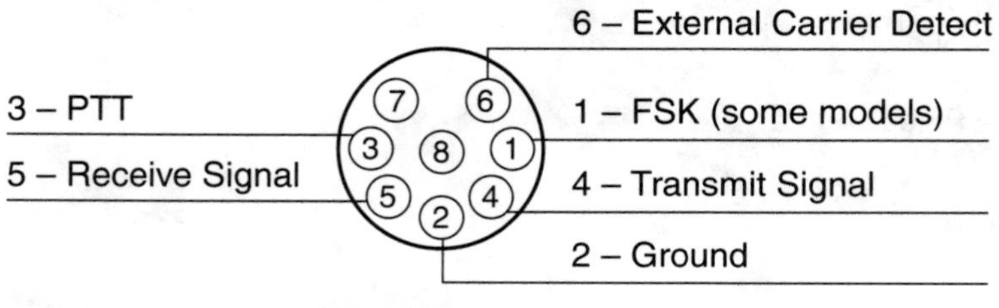

ICOM Radios with 13-pin accessory connector

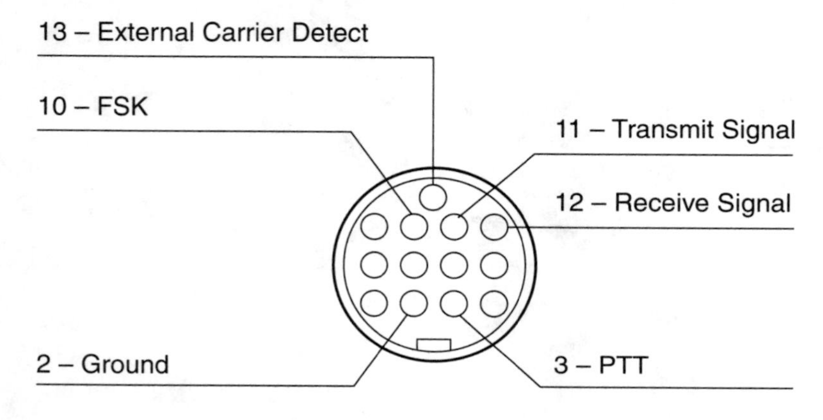

Notes: If you use FSK with the IC-706, see the manual for the setting the radio's tone, shift, and keying polarity. The pins on this connector are very close together and are very difficult to solder. To make it easier, remove the unneeded pins for the connector before soldering.

Kenwood 8-pin Modular (RJ-45) Mic connector

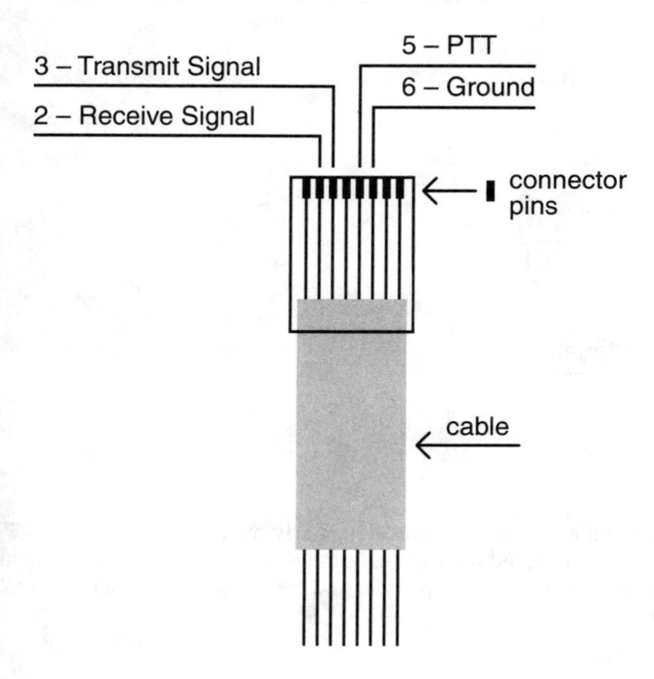

Notes: Pins are numbered 1 to 8 from left to right when looking at connector with the cable towards you, the locking mechanism down and the connector pins facing up. The Receive Signal is unsquelched. Many TNCs are defaulted to accept squelched audio. With unsquelched audio, the TNC will always be receiving noise and must be able to detect packet signals out of the noise. This is often accomplished with a TNC command for carrier detect.

Kenwood 13-pin DIN accessory connector, ACCY2

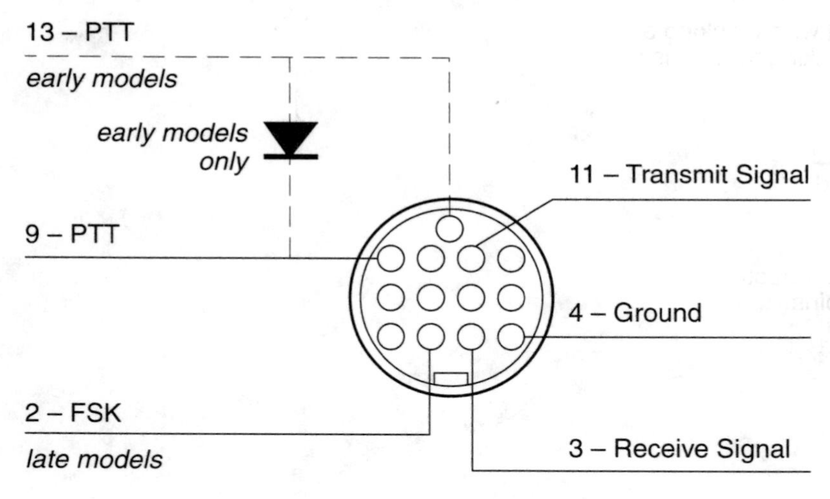

Notes: Pin 2 is used for FSK on late model radios. Later modes do not need Pin 13 connected. Early models have separate PTT pins for completing the circuit and for muting the mic. Both pins must be connected with a small-signal diode (1N914 or equivalent) as shown for early models. Check your manual. the pins on this connector are very close together and are difficult to solder. To make it easier, remove unneeded pins for the connector before soldering.

Kenwood 6-pin Mic connector

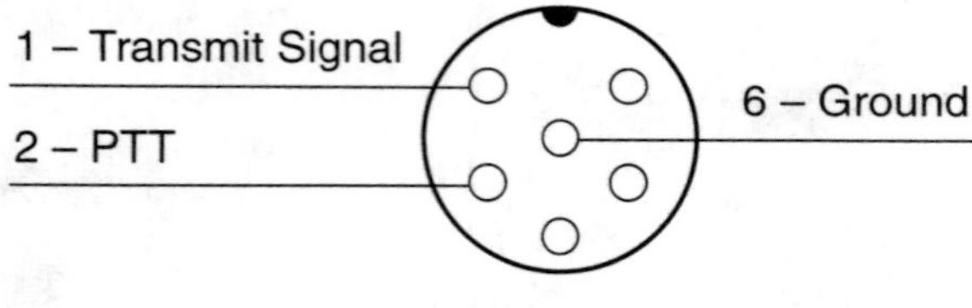

Kenwood Mic and Speaker plugs for the TH-78 and TR-2600

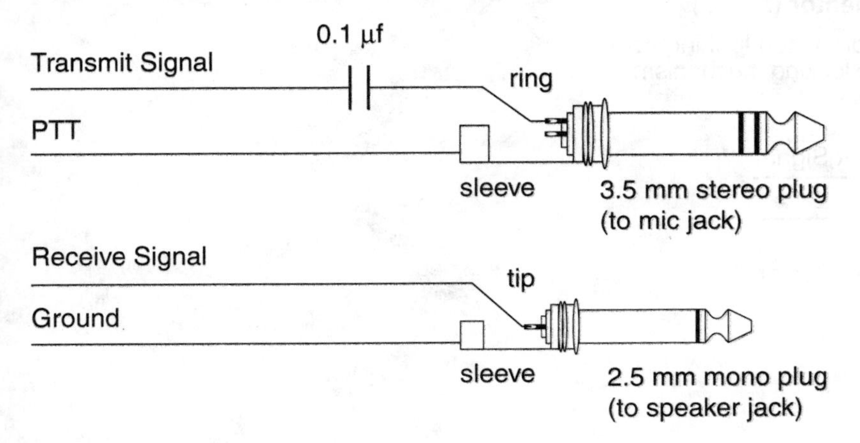

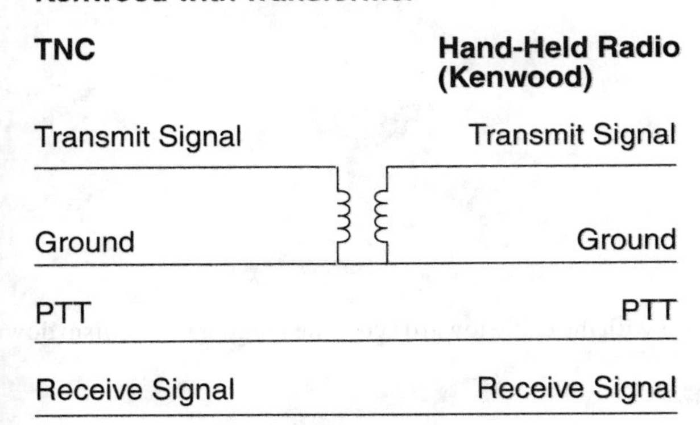

Notes: You may use a transformer instead of the blocking capacitor.

Radio Shack radios with 8-pin Modular mic connector

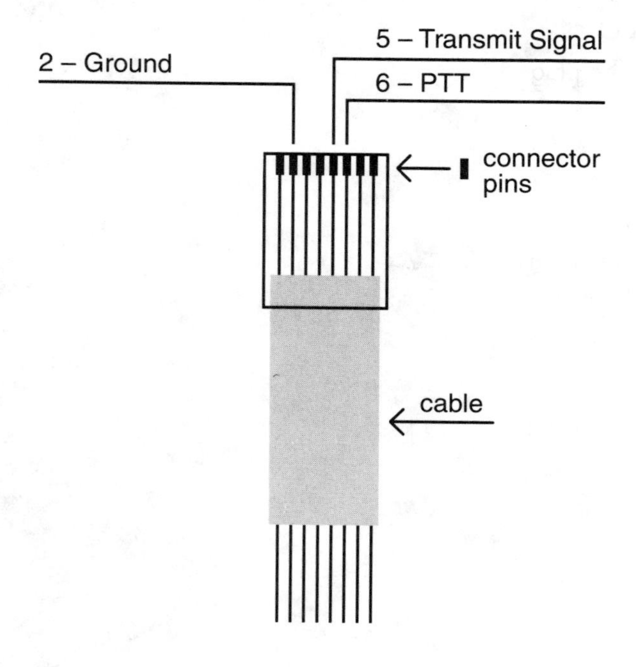

Notes: Pins are number 1 to 8 from left to right when looking at connector with the cable towards you, the locking mechanism down and the connector pin facing up.

Yaesu and Radio Shack Mic and Speaker plugs for FT-11, FT-530, FT-727, HTX-202, HX-404

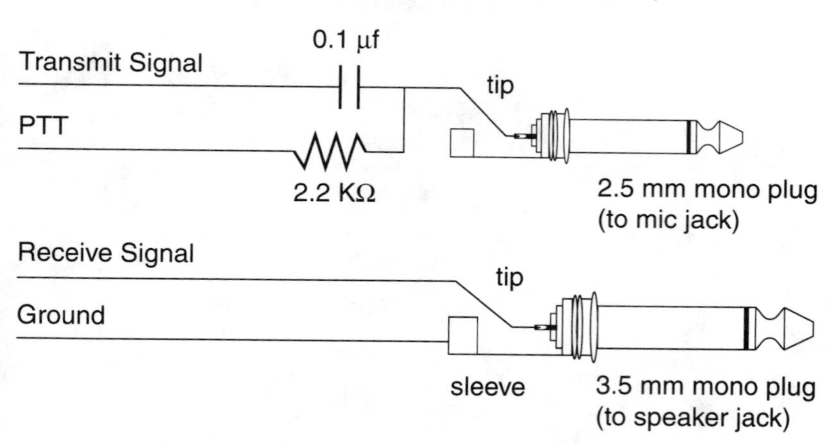

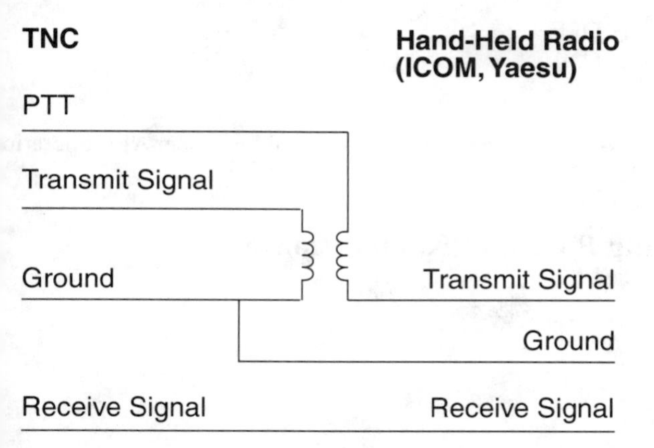

Notes: You may use a transformer instead of the blocking capacitor and resistor.

'aesu radio with 8-pin Mic connector

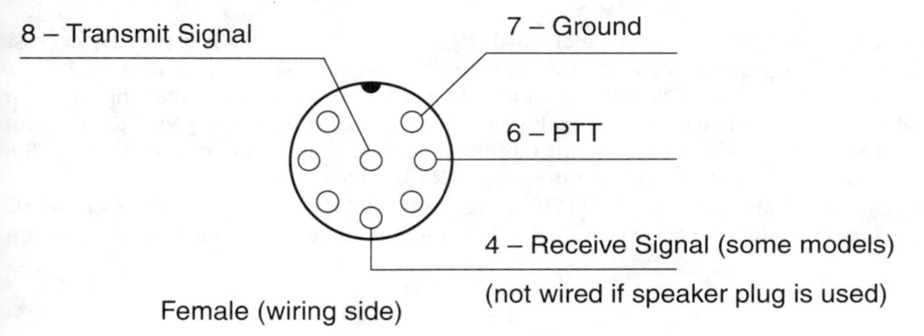

'aesu 5-pin DIN for FT-990 and FT-1000 when using AFSK

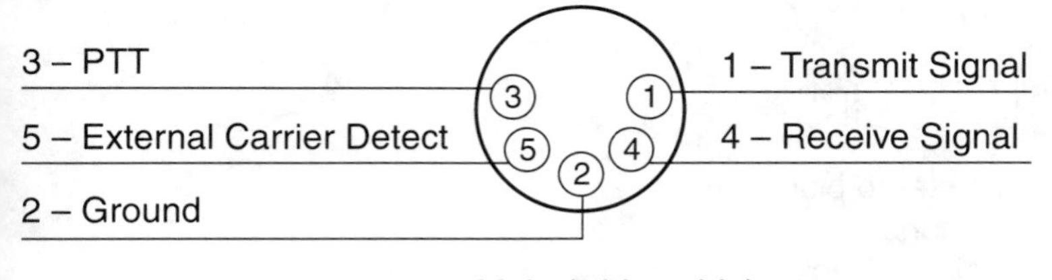

Male (wiring side)

Notes: The radio's Packet dip switches, which are located under the top access panel, need to be set to the tones used by the TNC. he radio's receive filters are centered on the tones selected with the Packet dip switches. Most TNCs use 2100/2310 Hz. For packet e KAM and KAM Plus use the tones set by their MARK and SPACE commands; for all other digital modes, they use the tones set y their MARK, SPACE and SHIFT commands.

The "Packet FSK Tone Pairs" chart in some Yaesu manuals is printed incorrectly. (Although this chart is called FSK tone, the ansmission method used is Audio or AFSK. The chart should read;

			TNC Tone Pair		
Dip Switches	1070/1270	1600/1800	2025/2225	2110/2310	
2		ON	ON	OFF	OFF
3		ON	OFF	ON	OFF

OFF is to the left and On is to the right when facing the radio.
This connector requires high level drive which can be adjusted in the TNC or with the radios Mic control for proper ALC operati

Yaesu 4-pin DIN for FT-990 and FT-1000 when using RTTY FSK operation

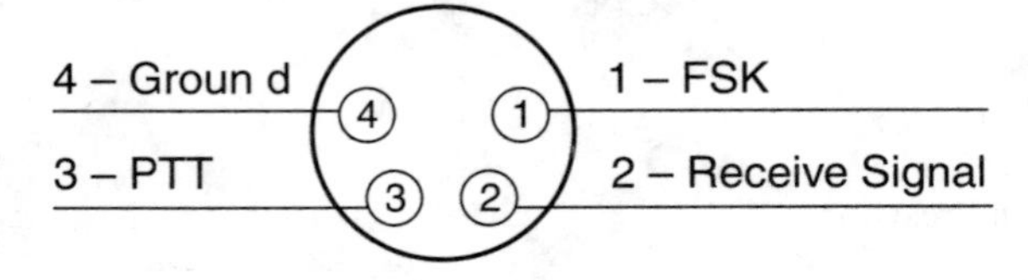

Notes: The radio's RTTY dip switches, which are located under the top access panel, need to be set for the shift and low to frequency used by the TNC. Shift is normally 170 Hz. Low tone frequency is normally 2125 Hz , except in Europe where 1275 is the norm.

The radio's RTTY Tone slide, which is located under the top access panel, needs to be set correctly. This is especially necess for AMTOR and RTTY/ASCII modes. Wiring the FSK pins between the radio and the TNC creates a simple contact-switch circu The TNC open can closes the switch. The radio's RTTY Tone slide tells the radio whether to transmit a high tone (space) or a l tone (mark) when the switch is opened (and the opposite when closed). Check your TNC manual to see if the TNC opens or clos the switch to send a mark. If the circuit is open when the TNC send a mark, the radio's Tone slide switch should be set to REV. So TNCs have an FSK invert command that can be used instead of changing the radio's RTTY Tone slide.

This connector may be used for any two-tone digital mode, such as AMTOR, G-tor, 300 baud Packet, Pactor, and RTTY/ASC When you select the RTTY mode (on the front of the radio), the radio uses the FSK pin on this connector to generate the transmit signal.

Yaesu Data IN/OUT plug for the FT-5100

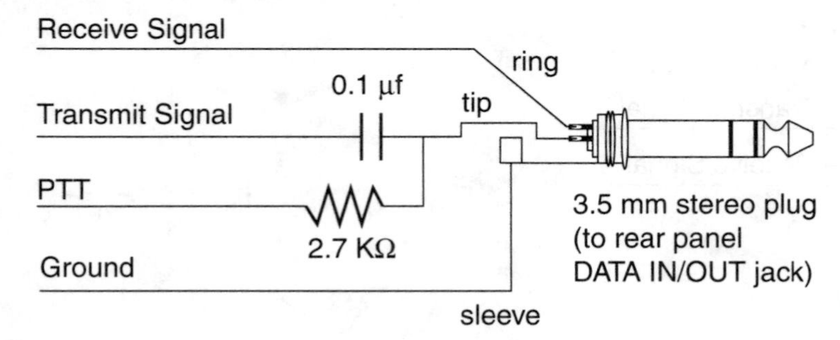

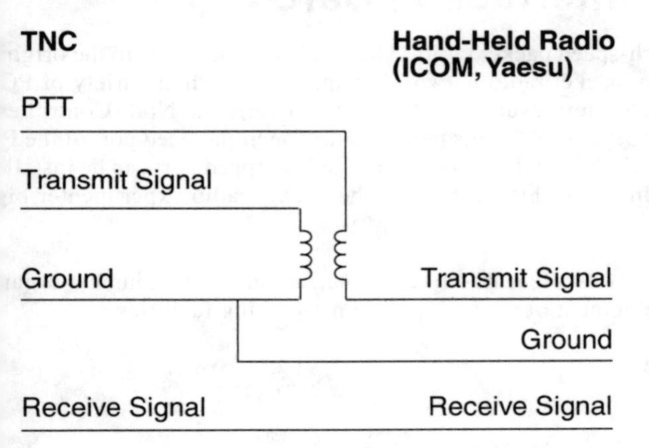

Notes: This data jack requires a high output drive level from the TNC. A transformer may be used instead of the resistor and capacitor.

Introducing the Ottawa PI2 (Packet Interface 2) Card

The PI2 card is a PC-compatible synchronous interface card for high-speed packet radio. The PI2 is the successor to the origir Ottawa PI card, first introduced in late 1990 and now widely used in packet radio networking applications, in a variety of PC ranging from XT's to 486's. The PI2 is a two-port interface card which offers a superior alternative to Terminal Node Controlle since it will not become obsolete when you wantto move up to speeds beyond 9600 bits per second. The high-speed port of the P is designed for operation at speeds up to at least 250 kilo bits per second. A 1200 bps modem for the low-speed port can be install on the card, allowing packet operation by simply adding a radio. In short, this card offers the packet radio experimenter hi performance at a reasonable cost.

The PI2 card is produced by the Packet Working Group of the Ottawa Amateur Radio Club, Inc., a nonprofit, volunteer orgar zation. All proceeds from the sale of the cards are used for the development of amateur packet networking facilities.

TECHNICAL SPECIFICATIONS

85C30 CMOS SCC chip (7.3 MHz clock)
Two half-duplex ports:
- The 'A' port: High-speed DMA
- The 'B' port: Low-speed interrupt-driven

High-speed port interface options:
- Buffered TTL (standard) for direct connection to modems, e.g., GRAPES 56 kb/s modem or Kantronics D4-10
- RS-422 (user-installed option)

Low-speed port is uncommitted. User can either install on-board TCM3105 1200bps modem, or install header for external TT connections
On-board modem has state machine DCD for 'open squelch' operation
High-speed port has provision for 32X bit rate clock output and rx data line gating, for G3RUH-type modems
On-board timer chip for reliable CPU-independent timing
Driver software, including source code, for KA9Q NOS included (backwards compatible with the original PI card)
Compatible with PC, PC/XT, and PC/AT type systems
Small size (6.25" length) and low power consumption o
Pre-configured software and detailed documentation included

The PI2 card and its predecessor have been extensively tested with the GRAPES 56 kb/s modem and KA9Q NOS, on various X and AT-class systems. Bench tests indicate that the high-speed port can support speeds of more than 460 kb/s.
Support is available via the Internet and Usenet.(Contact: **bm@hydra.carleton.ca**).

SUMMARY OF DIFFERENCES BETWEEN THE PI2 AND PI CARDS

A mechanical fit problem has been fixed (older PI card would not fit in some PCs unless the bracket was removed).
The PI2 has a CMOS SCC (85C30) instead of NMOS, and runs at twice the clock speed (7.3 vs 3.6 MHz). The CMOS SCC h several improvements over the NMOS chip, including relaxed timing constraints which should result in fewer problems related nonstandard PC bus timing. One result of the higher clock speed is that it extends the maximum bit rate for internal SCC clo recovery to 57.6 kb/s from 28.8 kb/s (note: maximum bit rate with external clocking is much higher than this).
The PI2 has a CMOS timer (82C54) instead of the NMOS 8253, and it runs at twice the clock rate. The PI2 high-speed port connections to the outside world now have TTL buffers (the PI only had direct connections to the SCC chip).
The PI2 high-speed port has provision for RS-422 interfacing (PI was TTL-only).
The PI2 can be populated with an on-board 1200 bps modem, including state machine DCD circuit and watchdog timer, driv by the low-speed port. The PI only had provision for external connections to this port (PI2 still has this too). A complete parts k for the modem is available.
The gate oscillator circuit in the PI has been replaced with a canned oscillator in the PI2.
The RX Data line is now qualified by DCD, to prevent unnecessary interrupts caused by noise.
The PI2 has provision for a 32X bit rate transmit clock output on the high-speed port, which can be used to run the state machi in G3RUH-type modems. The PI software driver has been modified to support this mode of clocking.
The PI software driver has been enhanced to allow different interface names to be attached to the low- and high-speed ports. T new driver is compatible with the old card.
The documentation has been extensively re-written and improved.
The PI2 is slightly larger (about 1.25" longer) than the PI.

ORDERING DETAILS

The PI2 card is available only fully assembled and tested (modem not included), for US $125, or CDN $155, plus shipping. complete parts kit for the on-board modem is available for US $30 or CDN $40 (documentation on the modem is included with eve PI2 card, so you can acquire your own parts if you wish). For orders from North America, shipping is by air parcel post. F destinations outside North America, shipping could be by surface mail. If you require faster delivery or shipping by other mean lease inquire before ordering.

The package includes KA9Q NOS (JNOS version) configured for the PI2 card, driver source, and installation documentation with chematic. In addition to the built-in driver for NOS, a packet driver (AX.25 class) compliant with the FTP Software Inc. specifiation is included in the package. The original PI card has been supported in NOS for several years now, and new releases of the rivers will be readily available from the usual Internet ftp sites (e.g., **ucsd.edu**, **ftp.ece.orst.edu**, and **hydra.carleton.ca**).

The low-speed modem kit contains a complete set of parts (including modem chip, crystal, and state machine ROM) and detailed nstallation instructions. Assembly and tune-up should take about an hour. The PI2 documentation also includes instructions on nstalling the RS-422 line drivers and receivers.

Send this order form to:

Attn: Packet Working Group
Ottawa Amateur Radio Club, Inc.
P.O. Box 8873
Ottawa, ON
K1G 3J2
CANADA

Name: ____Callsign: ____________
Address: ____
City: ________________________________Prov./State: ___________________
ZIP/Postal Code: _________________Country: ____________________________
Telephone: ______________E-mail address: ____________________________

Item	Quantity	Price	Total
PI2 card (assembled & tested):	______x	US$125 (CDN$155) =	____________
On-board 1200 bps modem kit:	______x	US$30 (CDN$40) =	____________
Add $10 (US$ or CDN$, flat rate per order) for shipping:			____________
Total amount enclosed:			____________

Please make money-order or check (sorry, no credit cards) payable to:
"Ottawa Amateur Radio Club".

MODIFICATION TO HELP WITH INTERMOD PROBLEMS WITH THE ALINCO 1200 DAT RADIO

CAUTION:

This radio employs static sensitive devices. Suitable means should be employed to assur a static-free environment prior to beginning work on this radio.

1. Remove power and antenna from rig
2. Remove bottom cover
3. Locate c-19, left side front of board (labeled, near q3).
4. Remove c-19: suck up the solder and carefully pry away from board while still warm.
5. Save the capacitor in case you got the wrong part!
6. Reassemble the radio, no retuning necessary.

This capacitor connects D1 (back to back diodes) to the first IF line. D1 is supposed to be a “limiter”. Unknown w Alinco chose to put in such a limiter, as the integrated IF chip IC-1, an MC3357, has a built-in limiter. Also, the Kenwo TM-231 which uses nearly the same layout and circuit, doesn’t have such a circuit. While the intermod hasn’t gone aw entirely (I’m not done with it yet!), it has diminished enough to make the rig usable on packet AND voice now.
73 de Bill, K0ZL@W0LJF.#NECO.CO.NOAM

ALINCO DR-1200 CONSTANT TNC RX AUDIO MOD

Pin 6 on the DR-1200 is connected to the speaker audio lead inside the rig to allow you to use a single cable to connect your TNC. The problem with this is that you have to leave th audio level up a little for the TNC to decode properly. Most people can only stand so much packet racket! Here’s a solution:

CAUTION:

The following procedure involves desoldering smd (surface mount devices). If you are not qualified to work on smd devices do not have the proper equipment, do not attempt this mod! Seek professional technical assistance. The originator of this message is not responsible for any loss, involving the performance of this modification.

Caution: this radio employs static sensitive devices. Suitable means should be employed t assure a static-free environment prior to beginning work on this radio.

1. Remove power and antenna from the DR-1200
2. Remove both top and bottom covers
3. Remove vol, sq, and vfo knobs

4. Remove vfo and mic plug nuts. Use needle nose as spanner, being carefully not to damage the threads.
5. Carefully remove the front plastic cover. Pry the four tangs gently!
6. Remove the three small phillips screws holding the control board to the chassis. Gently swing control board up to expose the back of the board.
7. Locate the pink wire connecting pin 6 of the mic jack on the control unit to the underside of the main board. Desolder the pink wire from the >main< board end and resolder that end of the pink wire to the “hot” side of the volume control. Identify the volume control by looking at the front panel that you have removed. The hot side is the terminal on the volume control (there are three) that is closest to the cpu chip.
8. Reassemble the radio.

NOTES: Some TNCS have a low input impedance on the RX audio line. The KPC-1 is 10 ohms, the KPC-2 is 600 ohms, but the PK-232 is 10Kohms. The lower impedance TNCS may “kill” the receive audio when you plug them in. The cure is to remove the offending load resistor in the TNC! If you do this, leave one end of the resistor connected for the next owner to reinstall, if needed. If you run regular 8 ohm audio into a TNC that has had the load resistor cut out, and don’t have a speaker connected also, you may damage your transceiver’s audio output stage. Caveats aside, the benefit is that now you can turn the audio all the way down and still get good copy. And, you can turn it up to monitor the channel without reaching around and fumbling with a speaker plug!
73 de BILL, Golden, CO
K0ZL@W0LJF.#NECO.CO.NOAM

MODIFICATION TO EXPAND TUNING RANGE OF THE DR-1200T TO 130-170 MHZ.

CAUTION:
If you transmit outside the designed limits of 144-148 MHz and/or the mars freqs, you may be risking your final pa transistors! The originator of this message is not responsible for any loss or damage or loss of use or safety if you decide to attempt this modification of your equipment.

(Horrible that we have to put disclaimers on something like this, but there’s always one sue-happy jerk who refuses to be responsible for himself...sigh)

This simple mod will expand the 1200’s tuning coverage and allow you to copy weather radio traffic on 162 MHz. Be sure to follow the manual’s instructions on tuning, since you can end up with 350 and 880 MHz displayed!

1. Remove power and antenna from the DR-1200
2. Remove the top cover.
3. With the front of the radio facing you, you will see a loop of yellow wire on the left, just behind the front panel. Cut this wire and tape the ends.
4. Reassemble the radio.

The radio will now tune from 130-170 MHz.

Caution:
This radio is not intended for commercial use or authorized for use in other then the amateur, races or mars services! Don't be a bootlegger!

-73- AB5S

GMSK Data Products

AX384 & AX576

High Speed Packet Radio Controllers

Data/AX384/A - Provisional Information

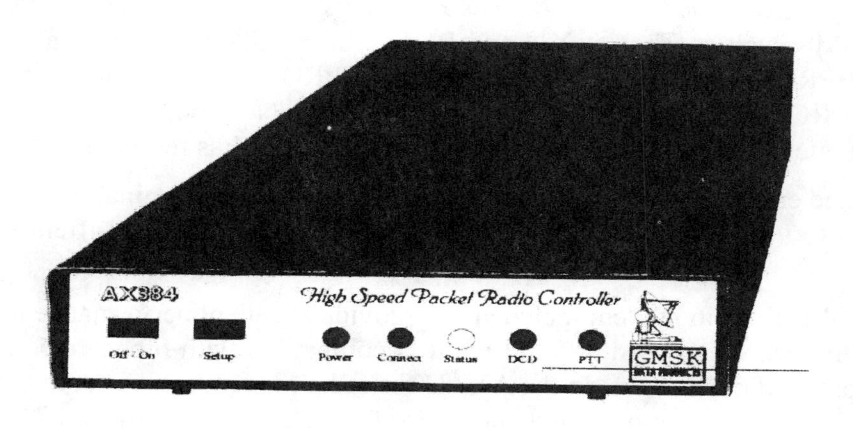

Feature List

- **Radio port** speeds of 4800, 9600, 19200 and 38400 b/s with AX384 or 7200, 14400, 28800, 57600 b/s with AX576
- **GMSK Radio Modem** operates full duplex at all the radio speeds above with no component changes
- **RS232 port** speeds from 9600 to 38400 b/s (AX384) or 115200 b/s (AX576)
- **TAPR TNC-2 EPROM** compatible including NET/ROM & 64k EPROM's such as TheNet X1J, and ROSE.
- **A Real Time Bit Repeater** can be enabled from software. This can provide a contention free LAN in its coverage area. An intelligent FIFO buffer is also included allowin transmissions of very long frames without bit under/over-runs. Other TNC functions are unaffected allowing simultaneous use e.g. as a Node - TheNet X1J
- **'Set-up' software** in ROM allows modem configuration to be modified on screen via simple terminal or enhanced Windows™ software program (supplied)
- **Live link Bit error rate measurements** enable easy setup of data links
- **Full Morse Ident**. as per UK. license regs., regardless of TNC software fitted. Can be disabled for use elsewhere !.
- **96k ROM space** allows in addition to 'Set-up and Kiss ROM', 1 x 32k EPROM, 1 x 64k EPROM or 2 x 32k EPROM images to be fitted.
- **128k RAM** can be fitted in place of normal 32k if required
- **Radio control signals** PTT & Mute can be set active high or low from Set-up software
- **10 Mhz Z80 Processor** ensures no lost or missed frames due to software errors.
- **Bi-Phase Data Coding** can be selected in place of data randomizer to allow simple interfacing to most types of voice radios (Includes FM / PM crystalled or synthesized)
- **RS485 Interface option** allows multiple TNC 's to be connected in 'Node Stack' with simple 4 wire cable. No more diode matrices !!

Description

The AX384 and AX576 are High performance packet data controllers designed for Amateur Radio users. Using the very latest in VLSI and RISC Processor technology these units set ne standards for Packet Radio TNC's. With radio operating speeds at a maximum of 38400 b/s or 57600 b/s respectively the AX384 and AX576 enable packet radio users to step up from the 1970's technology of 1200 b/s to the techniques of the 90's and beyond. The AX384 and AX576 are equally suited for use as High speed BBS ports, Network Nodes, or for the User that wants the very best from the packet radio network.

By maintaining software compatibility with the TAPR TNC-2 these TNC's support the most popular operating software such as TAPR v1.19, NET/ROM, and WA8DED 'Host Mode'. Even 64k EPROM's such as TheNet X1J and ROSE can be fitted with only a simple link change. 2, 32K EPROM images can be put into 1 64k EPROM and each run by simple link selection. These ROM's can be fitted in addition to the TNC operating ROM which is supplied. Both models can be fitted with either 32k or 128k RAM as required.

To ensure rapid and error free connections with the host computer or terminal the RS232 ports speeds can be set to 9600, 19200, and 38400 b/s. The AX576 additionally offers 57600 and 115200 b/s.

The full duplex GMSK radio modem included can provide excellent performance in the most demanding conditions. When used at 9600 b/s it is compatible with other modem designs, however it is capable of much more. In the AX384 the radio modem can be set to 4800, 9600, 19200 or 38400 b/s without component changes. For the AX576 the speeds are 7200, 14400, 28800 and 57600 b/s. Most types of high speed TNC use a data-randomizer developed by K9NG in their modem section. The AX384 and AX576 include this randomizer as a default option. However the user can also select an alternative method of data coding known as 'Bi-Phase coding'. The use of Bi-phase coding lowers the data rate possible in a channel but makes interfacing to typical voice radios much easier. These radios can by crystal controlled synthesized and FM or PM. When using 'Bi-phase coding' the data rates possible are 2400 b/s in 12.5 khz channels, or 4800 b/s in 25 khz for the AX384 or 3600 b/s and 7200 b/s for the AX576. In 'Bi-Phase coding' each data bit '1' is represented as '00' or '11' and a '0' is represented as '01' or '10', in this way there are no long strings of '1' or '0' sent over the air This ensures that the data can be sent in the normal speech audio band-width without excessive low frequency content.

A unique feature of the AX384 and AX576 is the inclusion of a 'Real Time Bit Repeater'. This allows the user to install a contention free LAN in an area very simply. The operation of the repeater is as follows. The repeating station is equipped with an AX384 or AX576 TNC and a Full Duplex radio. This radio has split transmit and receive frequencies. The stations wishing to use the repeater use split frequency half duplex radios. This is the same situation as with a voice repeater. As soon as one of the user stations starts to transmit data the repeater keys its transmitter and starts to relay the input data. As soon as the other user stations detect that the repeater is sending they are inhibited from transmitting. Thus all contention to access the repeater is avoided. Since the data is resent in real-time this is unlike a normal Node, where the packet must be fully received before it is forwarded to its end destination. The bit repeater in the AX384 and AX576 includes full data bit rate clock regeneration to avoid excessive clock jitter on the repeated data. It also includes an extending FIFO buffer to ensure bits cannot overrun or underrun where the bit rate clocks of the user stations and the repeater are slightly different.

The main operating software ROM supplied includes a 'KISS mode' driver for the TNC and a 'Set-up' program. For many users such as those running G8BPQ Node, KA9Q TCP/IP, and Linux™ ax25 software this is all that is required. The 'Set-up' program allows the user to simply configure some TNC and modem functions. The 'Set-up' program may be driven by a simple terminal program or a Windows™ based program (supplied with the TNC's). The screen photograph below shows the simple terminal based software.

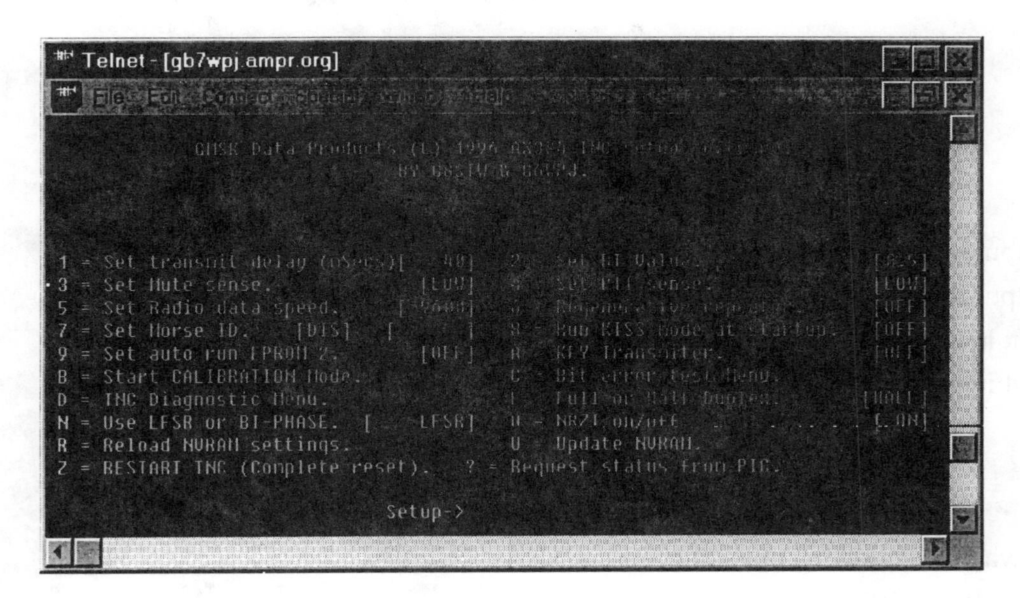

The 'Set-up' mode is activated by holding in the 'Set-up' button while switching on the TNC. It also allows various test signals to generated by the radio modem to aid the user get the best performance from the connected radio equipment. When used on a radio link with a AX384 or AX576 at each end the users may do direct bit error rate measurements on the link to allow 'tuning' of the RF equipment for best performance. The 'Set-up' program enables a on/off tone keyed Morse code identification signal to be sent every 29 minutes. This meets the needs of the UK license but can also be useful in situations of potential interference problems. The other features which can be configured by the 'Set-up' mode are shown in the 'screen shot' above. These include the sense of the radio interface signals PTT and 'Mute' which can be active high or low.

A 10MHz Z80 Processor is fitted as standard, This fast processor ensures there are no dropped or lost frames even at the highest operating speeds.

An optional RS485 interface allows Multiple TNC 'Node' stacks to be connected without diode matrices using simple twisted pair cables. This RS485 interface can be operated at 9600, 19200, 38400, and additionally on the AX576, 57600, and 115200 b/s, for the very best in multiple node stations.

As can be seen the AX384 and AX576 are 'state of the art' TNC's yet are available in both kit and built versions. Full user documentation is included in electronic form along with a Windows™ based program to control the 'Set-up' mode. The construction is from high quality materials including a RFI screen coated stylish grey casing. The styling is designed to match well with modern office and computer equipment. Finally the PSU is extensively filtered and smoothed to ensure correct operation of the TNC even in the presence of strong RF fields. The AX384 and AX576 are 'future proof' by design and construction !

Operating Specifications

Power Supply:			+9 to +15 volts via power connector
Current Consumption:			200mA (Typical maximum).
Radio Bit Rates:	AX384	-	4800, 9600, 19200, 38400 b/s
	AX576	-	7200, 14400, 28800, 57600 b/s

Radio Interface: Receive audio,
Transmit audio,
Press to Talk (PTT).
Mute Input.

Transmit audio output impedance	:	10k Ohms (max).
Transmit audio output level	:	0 Volts (min.) to 5 Volts (max).
Receive audio input level	:	50 mVolts (min.) to 5 Volts (max)
Receive audio input impedance	:	100k Ohms (min.).

Computer Interface

RS232C interface with the following operating speeds

AX384 - 9600, 19200, 38400 b/s

AX576 - 9600, 19200, 57600, 115200 b/s

RS485 interface also capable of working at the above speeds (optional)

Further Information

To find out more on these TNC products or other ax25 Packet Radio products, or to place an order please contact GMSK Data Products.

Thank you very much for your interest in the AX384 and AX576. Please watch out for further information on new GMSK Data Products developments for ax25 Amateur Packet Radio

GMSK Data Products
80 Colne Road,
Halstead,
Essex
CO9 2HP
England

Fax: +44-(0)-1787472290
Email: sales@nuthatch.dungeon.com

W3IWI/TAPR TAC-2 (Totally Accurate Clock) Project

March, 1997

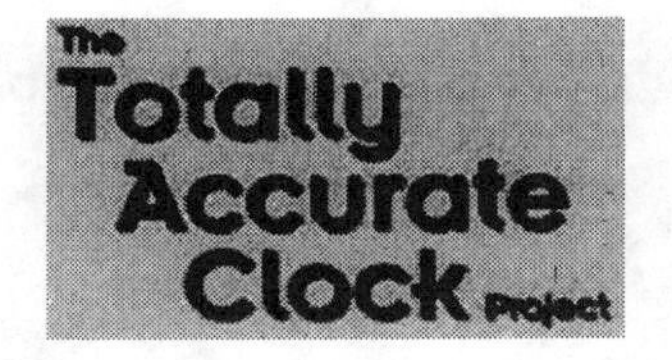

TAPR is actively working with Tom Clark, W3IWI, in developing the TAC-2 kit. The TAC-2 is the follow-on to Tom's original TAC project. The goal is to have a finished kit complete with parts and documentation in second quarter of 1997. Information on the TAC-2 project's progression will be documented here and on the TAPR-BB listserver.

Project Status

March 15th, 1997

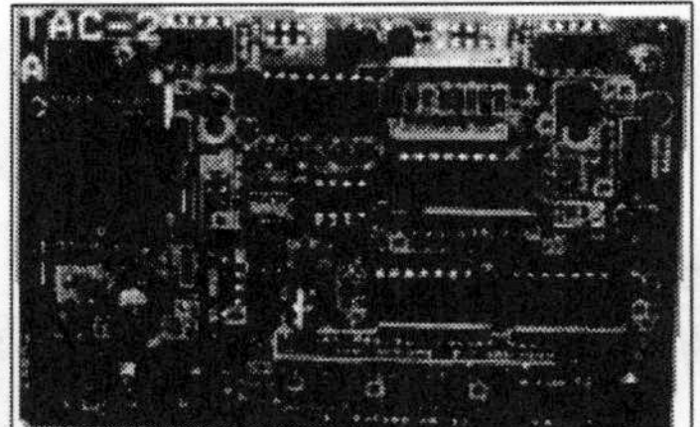

Testing continues and a final production board is nearly completed. Documentation is being completed and will be reviewed by between 3 and 5 people. More information will be presented when the final production kit cost and availability is know. The kit is looking very good. The pictures to the right are (top - component side of the TAC-2) and (underneath - Side view of the TAC-2 with Motorola Oncore-VP mounted underneath).

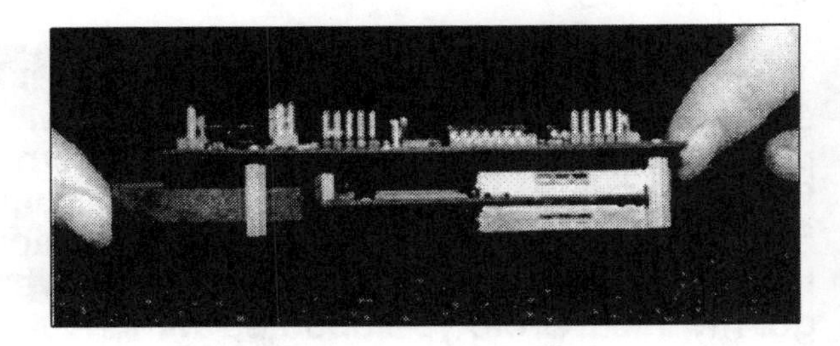

January 1997

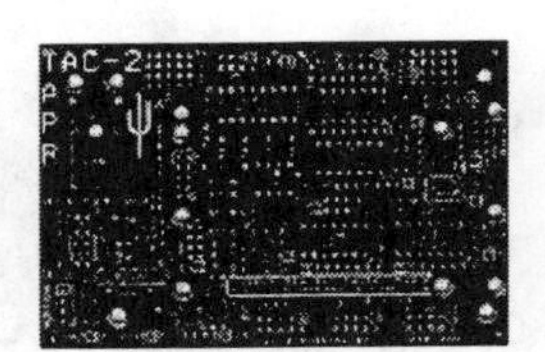

The project team has the rev-b boards in hand and are building. When docs are written and testing is completed with this round, we will be asking for Beta testers. Details on how to particpate in the beta-testing will be announced in the Winter TAPR PSR (Packet Status Register) and on the TAPR-BB e-mail list. It is expected that a small group of 5-10 people will be asked to particpate in the beta-testing.

November 1996

The project team is reviewing the design of the second alpha board. The second alpha-board will be put into the board shop (3-4 week turn) once everyone feels that it is ready. It is not expected that a beta run will be done until the Spring.

October 1996

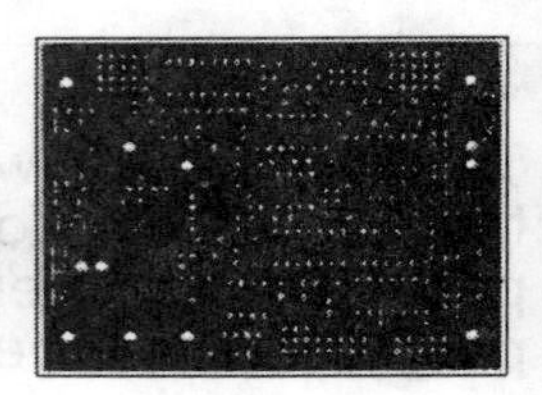

The first alpha boards arrive. Alpha testers installed components and found that they work! Documentation begins. Project teams decides to add functionality and additional components to board. This requires a change in the board layout and a second alpha board run.

September 1996

Purchase of the alpha PCB authorized by the TAPR BoD.

Totally Accurate Clock (TAC)

Excerpts from TOTALLY ACCURATE CLOCK ANNOUNCEMENT, Tom Clark, NASA/GSFC (February 2, 1995)

The "TAC" name is supposed to invoke a smile on your face. Many of you remember Heathkit's "Most Accurate Clock" (a WWV receiver) and I see advertisements for VLF clocks (WWVB in USA, DCF77 in Europe) that still use the "Most Accurate Clock" name in their advertising. Since the "TAC" is 3-4 orders of magnitude better than the "Most Accurate Clock" units, the "Totally" name seems warranted (also, TAC are my initials and this was begun as a home project!).

The TAC project began when I was on sabbatical at Onsala when Bernt Ronnang got me an early prototype of a Motorola PVT-6 OEM GPS receiver. In that incarnation, the PVT-6 was pretty disappointing. When I got back home, I had Motorola update the internal firmware and found that its personality had changed completely -- it was now very precise, but it had about a 500 nsec bias. I contacted a friend at Motorola who was involved in the PVT-6 software and he told me that tests at USNO had uncovered the same error. I was added to the "beta" group, got my initial prototype updated with the latest firmware and began more detailed testing. What I then found was that the PVT-6 receiver had the best timing performance I have seen in any small GPS receiver. With a small amount of care in setting it up, it now gives 50 nsec or better RMS timing precision and biases appear to be < 20 nsec.

The TAC project now involves both hardware and software. Let me briefly describe both to you.

HARDWARE:

The core of the TAC consists of a GPS. Several are going to be supported in the TAC-2 design (Motorola ONCORE, Garmin GPS-20, Trimble SK-8) The circuit board allows these various GPS to be mounted to the board.

The TAC-2 adds a number of desirable features:

- The 1 PPS output signals are buffered through a 74AC04 gate to improve the drive capabilities and to act as a "fuse" to prevent damage to the receiver in case of an operator goof. Up to three independent buffered outputs are provided, and the buffers will drive about +2 to +2.5 volts into a 50 ohm termination. The normal logic polarity is positive going at the epoch time, but this can be inverted if desired by some simple jumpers.

- Up to three open collector 1PPS signals are also available, negative going. Normally these would be used to drive display LEDs, but they can be used for other purposes.

- The add-on board includes an RS232 driver that provides 1 PPS time synchronization to an attached computer. The 1 PPS signal is normally connected to the computer's DCD input.

- The RS232 I/O to the computer is buffered and isolated from the receiver to act as a "fuse" to prevent damage to the expensive receiver in case of an operator goof. An RS232 OR-gate is supplied for the receiver input signal to allow RTCM SC104 Differential GPS signals to be fed to the GPS receiver.

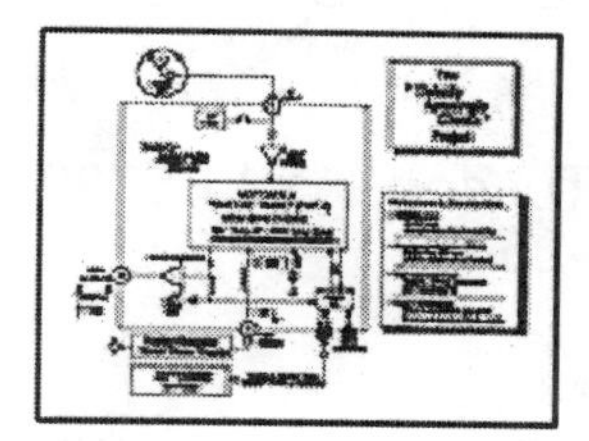

SOFTWARE:

A program called SHOWTIME displays the current time in BIG DIGITS you can see from across the room. In addition to just showing the UTC time, it includes a display of the date, day-of-week, day-of-year, local and Greenwich Mean Sidereal times, JD and MJD, and even the current GPS week. You can enable audible "WWV-like" time ticks to assist you in setting the formatter (or your wrist-watch). You can have the software automatically reset the PC's internal clock with about 25 msec accuracy. All the time display updates and audible ticks happen synchronously with the GPS 1 PPS signal because the PC reads the tick on its DCD line.

SHOWTIME allows you enter timing offsets and handles all the arithmetic for you. It allows you to make easy corrections for time delays in cables and the instrument and it tells you (with 1 nsec resolution) the actual epoch of the 1PPS tick and it gives you an estimate of the accuracy of the tick. SHOWTIME gives you a nice display of which satellites you using and which satellites are above the horizon. This includes a bar-graph "S- meter" for each of the GPS satellites currently in lock which are updated once per second.

The software lets you change operating modes (timing vs. position, elevation masks, satellite selection criteria, etc) easily and when you are running in position-determination mode, it will handle all the position averaging tasks for you. At any time, you can save the current configuration (positions, timing offsets, receiver modes, etc) to a disk file and restore that configuration at a later time.

At this time SHOWTIME runs stand-alone on a separate MS-DOS PC (but it does seem to run OK in the DesqView multitasker). Once you have set parameters into the GPS receiver, the PC operations can be terminated.

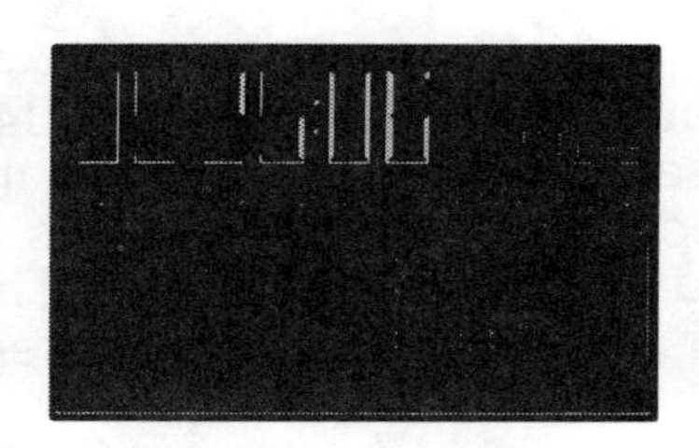

Visit Tom's TAC site

at: ftp://aleph.gsfc.nasa.gov/GPS/totally.accurate.clock

or

ftp://bootes.gsfc.nasa.gov/GPS/totally.accurate.clock/

Tucson Amateur Packet Radio
8987-309 E Tanque Verde Rd #337 * Tucson, Az * 85732
Office (817) 383-0000 * Fax (817) 566-2544
Internet e-mail: tapr@tapr.org

Greg Jones, WD5IVD, wd5ivd@tapr.org **(www)**

(This page last modified on March 13, 1997)

Emulating the "Totally Accurate Clock"
Tom Clark, W3IWI
April 14, 1996

-=-

A number of people have expressed an interest in making clones of my "Totally Accurate Clock" (TAC). This note will document ways you can fabricate a simple, but functional version of the TAC with little more than an extra wire in the cable.

-=-

HARDWARE DESCRIPTION:
The core of the TAC is a commercial OEM GPS receiver board. The boards that are supported are

1. Motorola ONCORE BASIC and BASIC EVAL models
2. Motorola ONCORE VP and VP EVAL
3. Motorola ONCORE XT
4. Garmin GPS-20 (and probably GPS-25, but it hasn't been tested)

To function as a TAC, I require that the receiver has

a. Computer serial I/O - at least with NMEA messages at 4800 baud (and for the ONCORE receivers, Motorola's proprietary binary protocol at 9600 baud).
b. A 1 Pulse per second output nominally synchronized with the UTC second. For the Motorola receivers, this means that either Option A (1 PPS Output) or Option I (RAIM + 1PPS) must be installed; the EVAL developer's kits have these options standard. The Garmin GPS-20 receiver has 1 PPS as standard feature. The Motorola 1 PPS signal is 0.2 seconds in duration and the Garmin 1 PPS is 0.1 seconds. Both have "TTL" level 1 PPS signals which rise zero to ~+5 volts at the nominal timing epoch.

The TAC project uses stock receivers with one important addition - the 1 PPS signal is available for precise epoch timing by the user (usually on a BNC connector) and it is also sent to the attached computer. I have adopted the convention that the 1 PPS signal to the computer will be on the RS232 DCD handshaking line. The TAC-to-computer interface requires 4 wires (preferably shielded):

FUNCTION	RS232 with 25-pin connector	RS232 with with 9-pin connector
GPS TXD from GPS	pin #3	pin #2
GPS RXD to GPS	pin #2	pin #3
1PPS from GPS (DCD)	pin #8	pin #1
Signal Ground	pin #7	pin #5
Cable Shield	pin #1	pin #5

The GPS receivers also require the user to provide external DC power:

* The Motorola ONCORE BASIC and XT (and the VP EVAL models in a metal box) all have DC-to-DC converters and power regulators so that the user can supply any supply voltage between +10VDC and about +30VDC. In this note I will call this +12VDC.

* The ONCORE VP and Garmin receivers require the user to provide regulated +5VDC power and will be damaged if the wrong voltage is applied. This can be easily generated by a 7805-type regulator chip.

All the GPS receivers have a multi-pin connector to supply power, provide computer I/O connections, and deliver the 1 PPS signal to the user. The Motorola BASIC and VP receivers use a 2x5 10-pin connector with pins on 0.1" centers; the user is warned that the BASIC and VP series of receivers use different connections and a receiver may be damaged if wiring for the wrong receiver type is used! The Motorola XT and VP EVAL receivers are housed in an extruded aluminum box and has a DB-9 connector. The Garmin receivers use a single-row 12-pin connector.

In addition to the power+I/O connector, all the receivers have a coaxial cable connector for an external antenna. +5VDC power is supplied on this connector for an antenna mounted RF preamplifier. All the receivers use a small MCX-series connectors, except for the metal-cased Motorola XT/VP EVAL models which use a BNC connector.

All the receivers except for the Motorola VP have RS232 voltage levels for their TXD/RXD serial I/O. The RS232 signals 'idle' at a voltage between -3 and -10 volts. They go to a voltage between +3 and +10 volts when a data "one" is sent. The Motorola VP uses 'inverted TTL' levels, idling at +5V and going to zero with data. Thus normal computer RS232 voltages must be inverted and level shifted if you are using a VP-series board-level receiver.

Motorola sells a developer's version of the VP receiver called the VP EVAL. The VP EVAL version of the VP has an internal circuit board that provides the RS232 level conversion and a +12-to-+5 volt DC-to-DC power converter. The same mechanical package is sold by Motorola as the XT model, except that the XT apparently uses the older ONCORE BASIC board.

-=-

SYNCHRONIZING THE COMPUTER:
The receivers all transmit time, date and position data to an attached com puter via the RS232 link. For simplicity I have chosen to use the NMEA-0183 (National Marine Electronics Association) standardized messages at 4800 baud. All GPS receivers have to develop precise timing in order to perform their navigation tasks. Each second they perform a lot of computations and then output the time and navigation results to the user. However, the user report is the lowest priority task, so the time-tag contained in the ASCII text in the NMEA message lags true UTC (Universal Time Coordinated) time by a fraction of a second. This has the effect of the voice announcement on WWV/CHU/JJY radio being like "At the tone the time WAS xx:xx:xx". Some users of GPS receivers have tried to determine (a real kludge, in my opinion!) ad hoc "lateness" errors so that they could obtain more accurate time.

I chose to adopt a more rigorous scheme to achieve computer synchronization. I determined that the timing of the 1 PPS signal generated on the receiver boards was very good -- for the Motorola receivers I can obtain 20-50 nano-second (nsec) accuracy and precision if I set up the receiver properly, which was the genesis of the TAC project. The Garmin GPS-20 is less accurate with ~1 microsecond (usec) performance -- still very impressive.

To achieve computer synchronization, I decided to send the hardware-generated 1 PPS signal to the computer using the Carrier Detect (DCD) RS232 handshaking wire. In software I would get the date/time from the NMEA test message, increment to the next second, and then wait for the DCD hardware signal to arrive. As the 1 PPS signal is received, the software applies the time that was calculated as appropriate to the next second. Thus the computer's time is accurate to the level of the latency of the recognition of the DCD hardware 1 PPS signal.

Although the RS232 specifications call for signal levels of -3 to -10v (low) and +3 to +10v (high), I have found that most modern PCs work reliably when driven with 'TTL level' signals (0v and +3 to +5v).

-=-

THE 'REAL' TAC DESIGN:
My initial TAC hardware design was designed around the original Motorola ONCORE BASIC (then called PVT-6) receiver. I designed a small circuit board to provide a number of interface tasks:

* Isolated, fast rise-time buffers to provide clean 1 PPS signals to the user
* RS232 line driver for the DCD 1PPS signal
* RS232 isolators to protect the receiver from user errors ('idiot fuses')
* An L-band preamplifier to improve receiver RF performance (and provide an 'RF idiot fuse')

* The ability to re-bias the antenna-mounted preamplifier with a voltage other than +5V
* Battery backup for the receiver's time-of-day clock and BBRAM

At NASA/GSFC we produced more than 50 such TACs which are in daily use at geodetic and astronomical observatories around the world. Their performance is documented in the many data files on my ftp server at the URL

ftp://aleph.gsfc.nasa.gov/GPS/totally.accurate.clock/

On aleph you will find a block diagram of the TACs we produced in the file

ftp://aleph.gsfc.nasa.gov/GPS/totally.accurate.clock/tac-blok.gif

and you will find a photo of the TAC at Onsala Space Observatory (Sweden) in

ftp://aleph.gsfc.nasa.gov/GPS/totally.accurate.clock/tac-foto.gif

The original Motorola ONCORE BASIC receiver used in the TACs is still available, but Motorola is phasing it out for the lower cost and higher performance (with 8 instead of the original 6 channels) ONCORE VP.

In addition to the Motorola receivers, I have developed the first prototype of the Garmin-based "TAC Lite" as a lower cost alternative for the radio amateur community. I am now developing a new TAC interface board that will support any of these receivers, but it is not yet available. It will probably be made available thru the Tucson Amateur Packet Radio (TAPR) amateur R&D group. TAPR has already announced the availability of the Garmin GPS-20 receivers for the TAC Lite design. Information on the TAPR-related developments can be found on TAPR's WWW page at URL:

http://www.tapr.org/

My aleph file server also serves as the repository for TAC support software and documentation. In particular, you may want to fetch the most current version of my SHOWTIME controller/display software and its documentation. With version 3.10 of SHOWTIME is included minimalist support for the Garmin GPS-20 'TAC Lite'.

-=-

TAC HARDWARE EMULATION:
Until the new TAC interface boards are available, I have drawn up some simple "TAC Emulation" sketches for each of the receivers is included in this emulate.zip distribution - the files are emulate.ps (in Postscript) and emulate.gif (in .gif format). The following discussion refers to these schematics.

All 4 versions shown in emulate.ps and emulate.gif have in common

* They assume you have +12VDC power
* The RS232 connections shown assume a standard IBM-PC 9-pin connector
* They all lack a 'proper' RS232 driver for the DCD 1 PPS connection and assume your PC will work with TTL levels.
* They show a 1 PPS BNC connector for your use in parallel with the DCD signal. The receiver's 1PPS output drive is limited, so the rise times are not as fast as I'd like to see.

An optional battery is shown. I highly recommend that you use a battery to help the receiver 'wake up' in a 'smart' mode. Some receiver boards already have a batter installed; if this is the case for you, then omit the option battery, diode and resistor. Each of the receivers seems to work OK with battery voltages from +3.5v to +9v. The NASA/GSFC produced TACs use a 9v NiCd rechargable battery and I have also used disposable 9v alkaline batteries without the resistor/diode charger. If your receiver has a battery built-in, it is probably a 3.6v unit.

If you need to provide a battery, you could use a 3.6v cordless telephone battery or a 9v NiCd battery (either obtained at Radio Shack). You will need a Silicon diode like a 1N4001 to prevent the batter from discharging when power is off.

You now need to compute the proper charge current limiting resistor. A fully charged NiCd cell ends up at a voltage of 1.4v. A "3.6v battery" has 3 cells, so its float voltage is 4.2v. A "9v battery" has 6 cells, so it floats at 8.4v. The diode will have a drop of ~0.6v across it. You will want to trickle-charge the battery with 5 to 10 ma. So for a "9v" battery being charged from 13.8v (not an untypical "12 volt" supply), the series battery charge current limiting resistor would be:

```
R = [(13.8v supply) - (8.4v battery) - (0.6v diode)] /(0.005 to 0.01 amps)
  =   480 to 960 ohms (I'd use 680 ohms)
```

or for a "3.6v" battery charging from +5V (out of the 7805 regulator):

```
R = [(5.0v supply) - (4.2v battery) - (0.6v diode)] /(0.005 to 0.01 amps)
  =  40 to 80 ohms (I'd use 68 ohms)
```

The most complicated of the emulator circuits is that for the ONCORE VP, since the RS232 logic levels must be inverted. In the new TAC interface board, I plan to use a MAX232/LT1181 RS232 level converter chip. Since the MAX232/LT1181 has two TX and two RX level converters, if you use it, you might as well buffer the DCD 1PPS line also. You could probably use a CMOS inverter chip (like a 74HC04) also.

-=-

Tom Clark

PACKET SPEED

PROOF OF PURCHASE

FEEDBACK

Please use this form to give us your comments on this book and what you'd like to see in future editions, or e-mail us at **pubsfdbk@arrl.org** (publications feedback).

Where did you purchase this book?

☐ From ARRL directly ☐ From an ARRL dealer

Is there a dealer who carries ARRL publications within:

☐ 5 miles ☐ 15 miles ☐ 30 miles of your location? ☐ Not sure.

License class:

☐ Novice ☐ Technician ☐ Technician Plus ☐ General ☐ Advanced ☐ Amateur Extra

---

Name ______________________ ARRL member? ☐ Yes ☐ No

Call Sign ______________________

Daytime Phone ( ) ______________________ Age ______________________

Address ______________________

City, State/Province, ZIP/Postal Code ______________________

If licensed, how long? ______________________

Other hobbies ______________________

____

Occupation ______________________

For ARRL use only	PKT SPD
Edition	2 3 4 5 6 7 8 9 10 11 12
Printing	1 2 3 4 5 6 7 8 9 10 11 12

From ______________________________________

____

____

Please affix postage. Post Office will not deliver without postage.

EDITOR, PACKET SPEED
AMERICAN RADIO RELAY LEAGUE
225 MAIN ST
NEWINGTON CT 06111-1494

please fold and tape